ASIC Design Implementation Process

Khosrow Golshan

ASIC Design Implementation Process

A Complete Framework

 Springer

Khosrow Golshan
Laguna Beach, CA, USA

ISBN 978-3-031-58655-2   ISBN 978-3-031-58653-8   (eBook)
https://doi.org/10.1007/978-3-031-58653-8

This Springer imprint is published by the registered company Springer Nature Switzerland AG
The registered company address is: Gewerbestrasse 11, 6330 Cham, Switzerland

If disposing of this product, please recycle the paper.

To those who are cancer survivors and those who are undergoing treatments, and to all medical professionals and those who work in essential businesses for risking their lives to save others.

Foreword

In the rapidly evolving landscape of semiconductor innovation, particularly at advanced nodes, the design of today's ASICs has become increasingly complex. The demand for quicker time-to-market has become an overarching imperative, especially in industries such as Internet-of-Things (IoT), automotive, biometrics, artificial intelligence, virtual reality, and 3D visualization.

This book not only acknowledges the pressing challenges but dives into the core issues encountered daily in the design of cutting-edge ASICs. With a focus on the ever-expanding domains where ASICs play a pivotal role, *ASIC Design Implementation Process—A Complete Framework* sets forth a thorough methodology, navigating the intricacies of area, power, and timing optimizations.

What sets this work apart is its commitment to offering a detailed walkthrough of each step of the design implementation process, from defining design requirements to ASIC design qualification, and everything in between. This is not merely a theoretical exploration but a hands-on technical reference book that provides real-world solutions to the challenges faced in contemporary ASIC design.

The appreciation for this book extends beyond its thoroughness; it lies in its ability to elucidate real problems and remedies throughout the design implementation process. Each chapter serves as a gateway to understanding the fundamentals with clarity, where engineers can find validation and enrichment in understanding how their expertise contributes to the bigger picture.

Managers and executives, often steering the strategic course of chip architecture, stand to gain great value from this work as well. The insights provided here facilitate more informed decision-making while also establishing a bridge of comprehension between the strategic goals of the executive realm and the day-to-day challenges encountered by the engineers in the trenches of ASIC design.

As you engage with this work, you will not only deepen your individual understanding, but you will gain an appreciation for the artistry as well as the collaborative brilliance required to craft cutting-edge ASICs.

Synaptics, Inc.
San Jose, CA, USA

Muna Moinuddin

Trademarks

Verilog, Encounter Design Integration, NanoRoute, Encounter System, Innovus, Conformal and CPF are registered trademarks of Cadence Design Systems, Inc.

Liberty, Prime Time, Formality, UPF and ICC are registered trademarks of Synopsys, Inc.

SDF and SPEF are trademarks of Open Verilog International.

DSP HDL Toolbox, DSP System Toolbox, HDL Coder, Simulink are registered trademarks of MATLAB, Inc.

All other brand or product names mentioned in this document are trademarks or registered trademarks of their respective companies or organizations.

Preface

This book aims to elucidate the process necessary for the advanced design implementation of Application Specific Integrated Circuits (ASIC). It offers understanding to some of the most challenging aspects of the ASIC design implementation process.

The field of ASIC design has evolved significantly since 1987, when a transistor gate length was approximately 3 um. Today, over three decades later, we are dealing with transistors with gate lengths of 5 nm and below, leading to increasingly complex ASIC designs. This evolution has made the signoff process for design implementation and time-to-market a considerable challenge. The trend of reducing transistor sizes and increasing design complexity is evident in modern Integrated Circuit (IC) process variations.

It is crucial to understand that a transistor's operating voltage is proportional to its gate length. For instance, at 3 um, the transistor's operating voltage would be 3.0 V, and at 0.18 um, it would be 1.8 V. As we move to lower nodes, the number of transistors increases by millions, making power consumption a critical factor. Smaller transistors have more leakage, and higher operating frequencies result in more dynamic power.

The 5 nm node, built using extreme ultraviolet lithography (EUV), was a significant milestone. It used a 13.5 nm wavelength to produce extremely fine patterns on the silicon. As we approach the limits of miniaturization, the question arises: what's next? The answer lies in reducing the Critical Dimension (CD) as much as possible, either by lowering EUV or increasing the Numerical Aperture (NA).

Moore's law has held true over the past few decades, with transistors becoming progressively smaller. Scientists and engineers have even attempted to create a one-atom-thick transistor. However, silicon remains the mainstream material for now. Given the current demand for smaller transistors and complex ASIC designs, understanding the importance of design implementation process is imperative. For instance, initial timing analysis on larger process nodes, such as 180 nm and 130 nm, were primarily concerned with operation at worst-case and best-case conditions. The distance between adjacent routing tracks was such that coupling capacitances were marginalized by ground and pin capacitance. As a result, potential issues with crosstalk and noise effects were often overlooked, leading to an increased design margin for analyzing crosstalk and noise.

Starting at 90 nm, and even more so at 65 nm, an increase in coupling capacitance due to narrower routing spaces and thicker metal segment profiles made crosstalk effects a significant concern. Different transistor voltages for speed (Low-VT) and leakage power reduction (High-VT) also became areas of concern. To address these issues, two additional corners were added: one to address the temperature effect (e.g., worst-case at $-40\,°C$) and the other to analyze the design leakage power (e.g., best-case at $125\,°C$). As process nodes became smaller (i.e., 40 nm and below), various aspects of the design implementation process were affected, including design requirements, validation, synthesis, physical design, verification, testing, and device qualification.

ASIC Design Implementation Process—A Complete Framework covers all aspects of design implementation. The topics covered include:

- Design Requirements
- Design Validation
- Design Synthesis
- Physical Design
- Design Verification
- ASIC Testing
- ASIC Qualification

Instead of delving into lengthy technical depths, the book emphasizes short, clear descriptions supplemented by references to authoritative manuscripts. The goal is to capture the essence of ASIC design implementation process for those involved in a particular part of the process and provide knowledge on other areas of ASIC design implementation process.

Laguna Beach, CA, USA
February 2024

Khosrow Golshan

Acknowledgments

Transforming a mere idea into a full-fledged book is as daunting as one might imagine. The journey is a rollercoaster of internal challenges and rewards.

Firstly, I extend my heartfelt gratitude to my wife, Maury Golshan. Her editorial expertise, unwavering dedication, and patience were instrumental in bringing this project to fruition. She invested substantial time and effort in proofreading and revising the manuscript, significantly enhancing its clarity and consistency. Without her relentless commitment, the completion of this book would have been nearly unattainable.

Furthermore, I wish to express my sincere appreciation to several individuals who generously contributed their time and effort to this manuscript. And I am grateful to Charles B. Glaser, the Editorial Director at Springer, and his esteemed colleagues for their invaluable assistance and support in the publication of this manuscript.

Khosrow Golshan

Disclaimer

The information contained in this manuscript is an original work of the author and intended for informational purposes only. The author disclaims any responsibility or liability associated with the content, the use, or any implementation created based upon such content.

Contents

About the Author

Khosrow Golshan was Division Director at Conexant System, Inc. and Technical Director at Synaptics, Inc. while managing and directing worldwide ASIC design implementation and standard cell and I/O library development for various silicon process nodes. Prior to that he was Group Technical Staff at Texas Instrument's R&D and Process Development Laboratory responsible for processing silicon test-chip design and digital/mixed-signal ASIC development. He has over 20 years of experience in ASIC design implementation methodology, flow development, and digital ASIC libraries design. He is the author of *Physical Design Essentials—An ASIC Design Implementation Perspective* and *The Art of Timing Closure—Advanced Design Implementation*. In addition, he has published many technical articles and has held several US patents. The author has earned degrees in the areas of Advanced Electrical Engineering (Southern Methodist University, Dallas, TX, Engineering Dept.), Master of Science in Electrical Engineering (West Coast University, Los Angeles, CA, Engineering Dept.), Master of Science in Applied Mathematics (Southern Methodist University, Dallas, TX, Mathematics Dept.), and a Bachelor of Science in Electronic Engineering (DeVry University, Dallas, TX, Engineering Dept.). He is also an IEEE life member.

Chapter 1
Design Requirements

Design is not just what it looks like and feels like. Design is how it works. –Steve Jobs

Requirements are the most important part of any Application Specific Integrated Circuit (ASIC) design. It defines the operation of an ASIC design. When defining requirements, it must meet its conformances. For example, if an ASIC operating at 50 MHz clock frequency exceeds its requirement of 40 MHz clock frequency, it is considered to have poor conformance. Thus, conformances are a measure of how well requirements are implemented during an ASIC design process.

A customer of a semiconductor firm is typically some other company who plans to use the ASIC in its systems or end products. So, the customer's requirements also play an important role in deciding how the ASIC should be designed. These ASIC providers are known as Original Manufacturer Equipment or OEM. As these companies increasingly turn to fables, ASIC provider and supply channel partners help build their complex ASIC designs. Thus, it's important to provide end-customers with high-quality designs that help them bring differentiated products to the market faster and with less risk. For that, the first step is to collect customer's requirements without any ambiguity, estimate the end product's market value, and evaluate the number of resources required to do the project.

Requirement capture is an essential part of any system's design. All the requirements must be captured so that the design can be implemented correctly. The implementation process of designing an ASIC is a transformation of one stage to another which goes from requirements, or concept, to an end-product (i.e., working silicon).

Because the end-product is typically extremely small (in mm^2), the design, implementation, and production processes are full of challenges and trade-offs, and the system architects, designers, and product engineers need to make engineering decisions which meet design requirements and conformances.

The engineering team need to clarify different steps in the ASIC design flow through well-defined quality assurance and control process starting from ASIC design requirements to design tape-out for manufacturing in the foundry and

© The Author(s), under exclusive license to Springer Nature Switzerland AG 2024
K. Golshan, *ASIC Design Implementation Process*,
https://doi.org/10.1007/978-3-031-58653-8_1

highlight important decisions and activities that each step entails. While the complexity of each step might depend on the choice of EDA vendor, the design application, and the technology node, the sequence largely remains the same.

Requirements are typically classified into types produced at different stages in a development progression, with the taxonomy depending on the overall model being used. For example, the following scheme was devised by Business Analysis Body of Knowledge [1].

Architectural requirements Architectural requirements explain what has to be done by identifying the necessary integration of systems structure and systems behavior, i.e., systems architecture of a system. In software engineering, they are called architecturally significant requirements, which are defined as those requirements that have a measurable impact on a software system's architecture.

Business requirements High-level statements of the goals, objectives, or needs of an organization. They usually describe opportunities that an organization wants to realize or problems that they want to solve and are often stated in a business case.

User (stakeholder) requirements Mid-level statements of the needs of a particular stakeholder or group of stakeholders. They usually describe how someone wants to interact with the intended solution, often acting as a mid-point between the high-level business requirements and more detailed solution requirements.

Functional (solution) requirements Usually detailed statements of capabilities, behavior, and information that the solution will need. Examples include formatting text, calculating a number, and modulating a signal. They are also sometimes known as *capabilities*.

Quality-of-service (non-functional) requirements Usually detailed statements of the conditions under which the solution must remain effective, qualities that the solution must have, or constraints within which it must operate. Examples include reliability, testability, maintainability, and availability. They are also known as characteristics and constraints.

Implementation (transition) requirements Usually, detailed statements of capabilities or behavior are required only to enable the transition from the current state of the enterprise to the desired future state but, after transition, will no longer be required. Examples include recruitment, role changes, education, and migration of data from one system to another.

Regulatory requirements Requirements defined by laws (federal, state, municipal, or regional), contracts (terms conditions), or policies (company, departmental, or project level).

It is essential that all the requirements are captured so that the design can be implemented correctly. Since there are several stages in complex ASIC design, each must be undertaken correctly because errors later in the implementation process become progressively more costly to correct.

Ideally the development process should incorporate all the required stages, and each one should be completed satisfactorily (requirements and specification) before moving on to the next. To meet requirements accurately, an initial study is needed for the ASIC project development engineer and marketing to work closely together.

During the initial study development and marketing, teams must identify a business opportunity and some initial product architectures which will satisfy the opportunity. The opportunity may arise because of a new market area or a new design to reduce costs or add new system features. Often, ASIC projects integrate several functions into one chip, reducing cost and power and typically increasing performance/power and providing additional features.

This is a key characteristic of the emerging SoC (System-on-Chip) market, where entire systems are integrated onto one chip (typically by reusing a combination of in-house and third-party Intellectual Property or IP). The deliverable from the initial study phase will be a business case that projects the financial returns, development timescales, and risks.

The development part of the business case needs to provide estimates for product costs, timelines (time-to-market), and resources needed. It should also provide a detailed assessment of the risks. For example, if the purpose of the ASIC is to replace a currently successful product, cost and/or feature enhancement is often the driving requirement. However, if the purpose of an ASIC is to address a new market area, timelines are often considered to have the highest priority.

During this initial study phase, the major building blocks of the ASIC must be identified with inclusion of preliminary engineering and manufacturing cost. It would be beneficial providing multiple development cost-based alternative product solutions and architectures, ASIC foundry, process nodes, external IP vs. internal design, etc. It's important to note cutting-edge products requiring advanced node processing technologies may require additional costs for adding new Electronic Design Automation (EDA) tools. From an ASIC design implementation perspective, the first requirements applicable to its design could be defined as:

- Specifications
- Architecture
- Initial design

The cost to detect and fix errors increases during ASIC design implementation stages. For example, if it would cost N units to fix a problem during the architectural phase, the same problem would cost more to be fixed depending on stage of design implementation. Figure 1.1 illustrates costs associated with an ASIC design implementation stage.

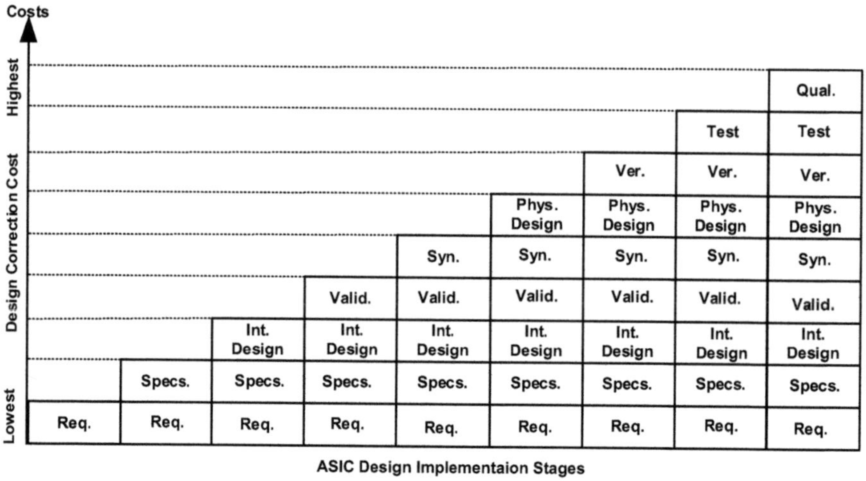

Costs

								Qual.
							Test	Test
						Ver.	Ver.	Ver.
					Phys. Design	Phys. Design	Phys. Design	Phys. Design
				Syn.	Syn.	Syn.	Syn.	Syn.
			Valid.	Valid.	Valid.	Valid.	Valid.	Valid.
		Int. Design	Int. Design	Int. Design	Int. Design	Int. Design	Int. Design	Int. Design
	Specs.	Specs.	Specs.	Specs.	Specs.	Specs.	Specs.	Specs.
Req.	Req.	Req.	Req.	Req.	Req.	Req.	Req.	Req.

ASIC Design Implementaion Stages

Fig. 1.1 Costs of an ASIC design at different implementation stages

1.1 Specifications

Developing a thorough and correct specification or design initiation usually sets a solid foundation for the complex ASIC design. The technical specifications need refinement of the technical requirements after initial design analysis, but it is important to cover the information in an unambiguous manner. In general, specifications describe the functionality, interface abstractly, and overall architecture of an ASIC to be designed.

An ASIC's well-thought-out design specification is particularly important because complex ASIC design cycles may be anywhere between a few months to a year. It is, therefore, important to foresee and predict what trends would be relevant a few years down the line if one needs to provide their product to a range of end-users.

As technology becomes more advanced and defined in every aspect of life, end-users expect new features and design improvements from their devices, including high-speed processing and low power consumption. A top-down design approach is employed to navigate and manage complexities of the complex ASIC design and its implementation process and, as a first step, dictates the development of a proper detailed specification. A thoroughly crafted working specification helps guide the design implementation process, with the project less prone to errors disruptive to project schedule and cost.

It is especially important that an expert in complex ASIC system design assists end-users in developing system architecture and specifications. This extensive process may take a few weeks or more depending upon the complexity of the application requirements. In general, the specification process may:

- Review multiple architectural options
- Analyze architecture options through technical feasibility
- Define functional requirements specification
- Reviewing the block diagram, system schematics, and specifications
- Develop an understanding related to design problems, operating environment, and challenges
- Decide components related to the final product, not only ASICs
- Determine if any certification is required in the product compliance
- Design or compile the ASIC block diagram with full functional components, specifications, and pin-out
- Decide board-level architectural trade-offs that lead to the most cost-effective silicon integration
- Generate the top-level architecture document
- Identify possible third-party IP block requirements
- Estimate silicon (chip) area, power, process, and cost

The specification can be viewed as an abstracted engineering form of design requirements and lists how a device needs to function and perform in various operational situations and is considered an extremely significant part of the complex ASIC design and implementation process. Once design specifications have been collected, they should describe the functionality, interface abstractly, and overall architecture of the complex ASIC to be designed. For example, Figure 1.2 illustrates a conceptional ASIC design for a graphic ASIC design.

The specification for graphic ASIC design as shown in Fig. 1.2 might have details such as:

- Computational power to execute imaging algorithms to support graphical display
- Vertical and horizontal pixel display numbers and its corresponding controller
- Macro Processor, Central Processing Unit (CPU), Digital Signal Processor (DSP) types and its operating speed
- Graphic Processor with Rendering Engine, Display Controller, Color Look Up Table (CLUT)
- Timing Controller (TCON)
- Automotive Pixel Link (APIX) that is a serial high-speed Gigabit Multichannel link used to interconnect displays, cameras, and control units over one single Transmit (TX) and Receive (RX) data
- The Advanced High-performance Bus (AHB) Bridge is used for connecting components that need higher bandwidth on a shared bus as well as connecting internal memories, Micro Processor, Graphic Processor, and/or external memory interfaces
- Double Data Rate (DDR) for synchronous dynamic random-access memory data transfer
- General Purpose Input/Output (GPIO)
- Universal Standard Bus (USB)
- Inter-IC Sound (I2S) for digital sound transfer

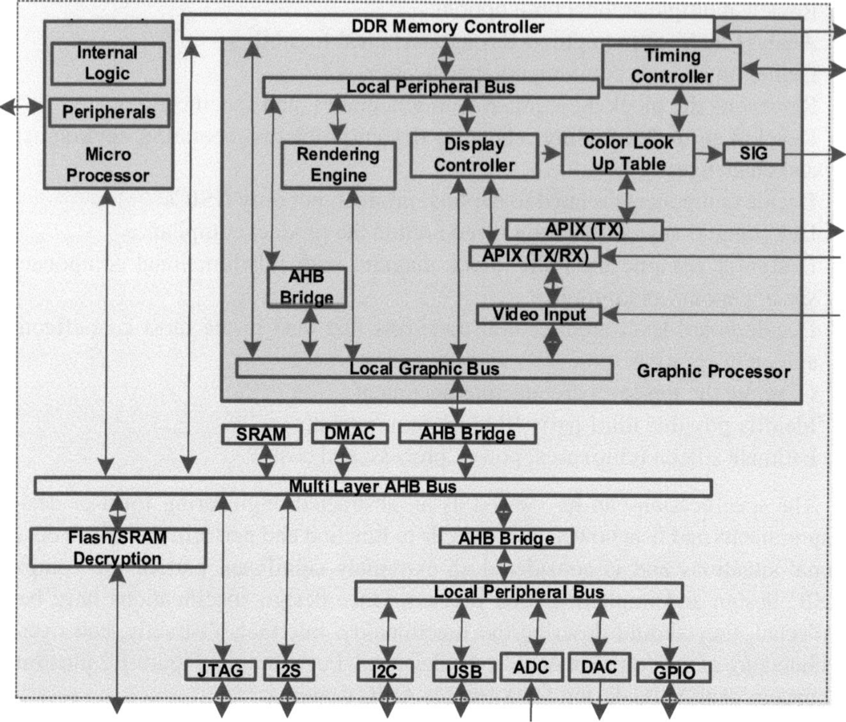

Fig. 1.2 Basic graphic ASIC design system-level architecture example

- Inter-Integrated Circuit (I2C) for attaching lower-speed peripheral to processor microcontroller for short distances
- Joint Test Action Group (JTAG) interface for verifying designs and testing printed circuit boards after manufacture
- Flash/Static Random Access (SDRM) memory for decryption
- Direct Memory Access Controller (DMAC)

1.2 Architecture

An ASIC architecture (e.g., graphic ASIC shown in Fig. 1.2) is a system-level view of how an ASIC design should operate. It provides all components required, what clock frequencies the system should run, and how to target power and performance requirements. At this stage, the entire ASIC functionality is divided into multiple functional blocks by analyzing possible options while considering performance implications and technical feasibility. In addition, it describes how the data should flow inside the ASIC. An example would be the data flow when the processor receives imaging data from the system memories and executes them and then graphics processors execute post-processed data from the previously retrieved data and store them into another part of memory.

During the architecture phase, the system-level view is decomposed and partitioned into smaller modules (macro architecture), and the interfaces between the modules are defined. Ideally, the hierarchy of the architecture should be captured diagrammatically, using a suitable graphical tool. The divide between hardware and software blocks is also a critical part of this phase of the ASIC design. Typically, the design is captured in a high-level programming language like C++ or System C.

Additionally, an architectural ASIC should be created using the Verilog Hardware Design Language (VHDL). The ASIC behavioral model should be simulated to verify that the desired functionality is captured completely and correctly. The behavioral abstraction is then used as a reference to create and refine an RTL abstraction that captures the desired functionality required by the design specification. It should be noted that, with VHDL, the difference between a purely behavioral and an RTL abstraction may appear subtle at this point, but it is important to have clear understanding of the distinction between them. Generally, an ASIC design represented in VHDL consists of three levels of abstraction:

- *The Behavioral Level:* A design is implemented in terms of the desired algorithm, much like software programming and without regard for actual hardware. As a result, a VHDL model written at the behavioral level usually cannot be synthesized into hardware by the synthesis tools.
- *The Register Transfer Level:* A design is implicitly modeled in terms of hardware registers and the combinational logic that exists between them to provide the desired data processing. The key feature is that an RTL level description can be translated into hardware by the synthesis tools. This is also known as Hardware Design Language (HDL).
- *The Structural Level:* A design is realized through explicit instances of logical primitives (e.g., AND/OR gates) and the interconnects between them. This is also referred to as a gate-level description or simply a netlist.

The ASIC design modeled in HDL is known as a pre-synthesis description and is used for simulations to verify that the RTL abstraction fully provides the desired functionality. The functional verification of the design that occurs at this point must be as complete and thorough as possible. This requires that the designer fully understands both the design specification and the RTL implementation. The test vectors used during simulation should provide the coverage necessary to ensure the design will meet specifications.

Using Fig. 1.2 as an example, the requirements dictate to incorporate DSP as part of its core logic. In this case, DSP takes real-world signals like voice, audio, video, and position that have been digitized and mathematically manipulates them (DSP is designed for quickly performing mathematical functions like "add," "subtract," "multiply," and "divide").

Signals need to be processed so that the information they contain can be displayed, analyzed, or converted to another type of signal that may be of use. In the real world, analog products detect signals, such as sound, and manipulate them.

Converters such as an Analog-to-Digital Converter (ADC) then take the real-world signal and turn it into the digital format of 1's and 0's. From there, the DSP takes over by capturing the digitized information and processing it. It then feeds the digitized information back for use in the real world. It does this in one of two ways, either digitally or in an analog format by going through a Digital-to-Analog Converter (DAC). All of this occurs at very high speeds.

An example of designing DSP for an ASIC Graphic Processing Unit (GPU) design using MATLAB involves several steps. Here are some key points:

- *Algorithm Development:* DSP design often starts with algorithm development and testing using MATLAB functions. MATLAB code, which usually operates on matrices of floating-point data, is good for developing mathematical algorithms, manipulating large data sets, and visualizing data.
- *Hardware Implementation:* MathWorks HDL products, such as DSP HDL Toolbox, allow you to start with a mathematical model, such as MATLAB code from DSP System Toolbox, and design a hardware implementation of that algorithm that is suitable for Field Programmable Gate Arrays (FPGAs) and ASICs.
- *Design Trade-offs:* Hardware designs require trade-offs of resource usage for clock speed and overall throughput. Usually, this trade-off means operating on streaming data and using some logic to control the storage and flow of data.
- *HDL Code Generation:* To bridge the gap between mathematical algorithm and hardware implementation, use the MATLAB algorithm model as a starting point for hardware implementation. Make incremental changes to the design to make it suitable for hardware, and progress toward a Simulink model that you can use to automatically generate HDL code by using HDL coder.
- *Cycle-Based Modeling:* While both MATLAB and Simulink support automatic generation of HDL code, you must construct your design with hardware requirements in mind, and Simulink is better suited for cycle-based modeling for hardware. It can represent parallel data paths and streaming data with control signals to manage the timing of the data stream.
- *Fixed-Point Type Choices:* Fixed-point type choices clearly visualize data type propagation in the design. It also allows for easy pipelining of mathematical operations to improve maximum clock frequency in hardware.
- *Golden Reference:* While creating hardware-ready design, use the MATLAB algorithm as a golden reference to verify that each version of the design still meets the mathematical requirements.

It should be noted that, since DSP applications are primarily algorithms that are implemented either on a DSP processor or in software, a fair amount of programming is required. Using interactive software, such as MATLAB, it is now possible to place more emphasis on learning new and difficult concepts than on programming algorithms. Figure 1.3 shows the concepts of a DSP design for a given an ASIC project.

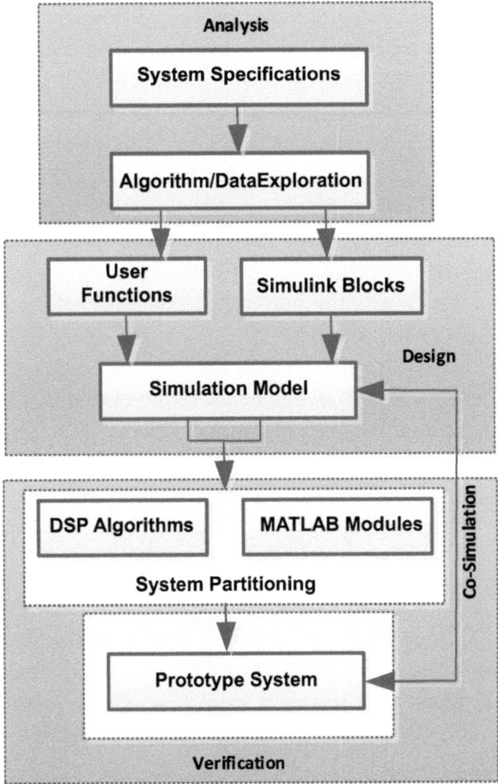

Fig. 1.3 The basic concept of DSP design

1.3   Initial Design

Initial design refers to transferring architectural ASIC design to a readable machine language, known as Register Transfer Language (RTL), in an abstract manner like a programing language (e.g., HDL) and creating a "pre-synthesis" model which can be interpreted by EDA tools correctly for design simulation and verification. The HDL model could be composed with multiplexing, decoding and case statements, etc. In general, an ASIC design pre-synthesis model, or netlist, often is comprised of the following:

- *Combinational Logic:* Combinational logic usually refers to Boolean combinatorial gates like the OR, AND, NAND, NOR, etc. While these gates are simple, they can be combined to perform complex digital operations.
- *Sequential Elements:* Sequential elements play a critical role in interfacing between different combinational logic clouds performing different functions by storing their output temporarily. These sequential elements, e.g., Flip-Flops and latches, are also referred to as memory elements and are controlled by a synchronizing or control signal referred to as a clock.

- *Finite State Machines (FSMs):* These are higher abstraction of a sequential logic which can be implemented both in hardware and software. FSMs model response of a digital machine to a set of inputs to produce deterministic set of outputs and serve as an important building block for logic designers.
- *Arithmetic Logic Blocks:* Arithmetic computations form the heart of the computing logic and usually are the bottlenecks for performance in high-performance CPU cores. Arithmetic computation includes addition, subtraction, multiplication, and division. There are numerous possible implementations of these circuits which offer a trade-off between performance, area, and power. Logic designers can choose the best suited computation for their application to optimize one or more parameters.
- *Data-path Design:* In addition to coding combinational logics and sequential elements, HDL can be used to model data path design in an abstract manner like a programming language which can be interpreted by EDA tools correctly. These models could be multiplexing, decoding, case statements, etc.
- *Analog Design:* In addition to digital logic, an ASIC may have many analog components to help in interfacing with the real world and may comprise of Temperature Sensors (TA), ADC, DAC, and, most importantly, the clock generating unit the Phase Locked Loops (PLL).

The following shows an HDL example of a D-type Flip-Flop for synchronous with clear. The output will reset at the triggered edge (positive edge in this case) of the clock after the clear input is activated.

```
module HDL-DFF (D,CLK, CLR, ,Q, NQ) ;
input           D, CLK, CLR ;
output reg  Q, NQ ;
always @(posedge CLK) ;
begin
    if (CLR==1)
      Q   <= 0 ;
      NQ <= 1 ;
    else
      Q <= D ;
      QN = !D ;
  end
endmodule
```

As mentioned earlier, the purpose of the initial design is to verify the ASIC architecture for a given product specification and prepare an abstracted design. For verification purposes, an initial design (i.e., structural model) coded in HDL should be checked to make sure that it functions as expected (functional verification) before validation and synthesis. Typically, verification goes in parallel with the ASIC design activity. During the initial design phase, hardware specifications and the verification plan are created, and test cases are built to see whether the structural

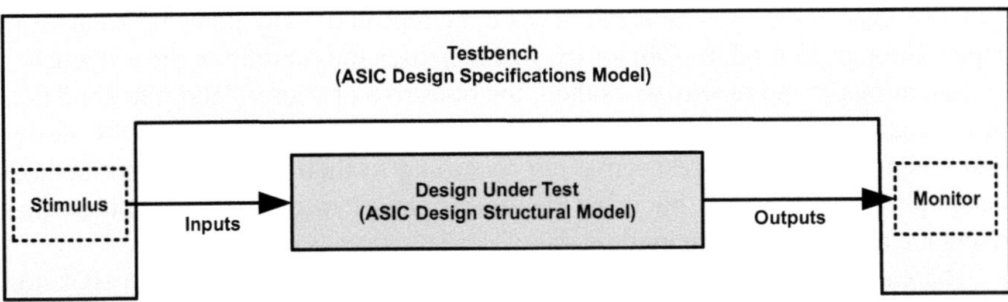

Fig. 1.4 Testbench basic structure

design architecture works correctly, or not, according to the design specifications. The test is then run on the RTL code to see if the design matches the specification. The design that is being checked is called Design Under Test (DUT), while the test that verifies, by applying stimulus and monitoring the output DUT, is called "testbench" as shown in Fig. 1.4 above:

Testbench needs to operate over a wide range of abstraction levels, creating sequences and transactions that are eventually translated into bit vectors. One of the main important features of the testbench should be maximum code reuse.

To verify a complex ASIC with many features, the testbench would contain hundreds of directed tests (i.e., stimuli and monitors). If constrained-random stimulus is used, far fewer tests will be needed. In general, constructing a testbench needs to contain all lower layers of the testbench such as:

- Scenarios
- Commands
- Functions

A testbench is specific for a DUT (ASIC design) and is coded in HDL as well. It would contain an empty `entity` and its `architecture` and must also have a defined `entity`. The entity describes the input and output of the circuit that are being tested.

Inside the architecture of testbench, there is a declared component which is DUT, or initial structural model of the ASIC design architecture, which needs to be initialized by signals. This is done to determine if values have been previously assigned. Then architecture would then use signals for internal calculations and assign the signal value to the port.

The next step is to generate a stimulus or any sequences for inputs of the initial ASIC design. Commonly, there are two ways to generate an in-program stimulus or inside of the testbench using:

- Repetitive pattern method
- Vectors pattern method

In the repetitive pattern method of stimulus generation, one bit of stimulus is generated and repeated multiple times. For example, one statement is coded to generate only one bit function and needs to be repeated many times for each set of inputs for a given module. Obviously, that increases the runtime of the testbench.

In contrast to the repetitive method, the objective of vector pattern method is to apply a group of stimulus bits to a port (a group of pins) in the initial ASIC design and assign the bit to its respective pin according to their position. Both methods accomplish the same task, but using the vectors pattern method is more concise than using the repetitive pattern method.

There are multiple methods for generating or coding the testbench in HDL format such as:

- Simple testbench
- Testbench with a process

The simple testbench and the testbench with a `process` type are more suitable for combinational circuits. For sequential circuits, the following types of testbench are more suitable.

- Infinite testbench
- Finite testbench

Additionally, infinite and finite testbenches are suitable for when there is a need to test something that is unpredictable (e.g., random sequence generator). It is important to note that the testbench code is used by all the tests which means it should be generic. To see how this architecture looks like, one could check the system Verilog environment examples. There are different tools for HDL verification by EDA companies such as Synopsys, Mentor Graphics, and Cadence. In general, when describing a design [2] using VHDL (or also known as System Verilog), one needs to declare an entity to define the inputs and outputs of a design as follows:

```
entity <entity-name> is
        generic ( <generic-name> : <type> ) ;
             port (
                <input_name>    : in <type> ;
                <output_name> : out <type) ;
                <bidir_name>     : inout <type>
                     ) ;
end entity <entity_name> ;
```

However, as the testbench has no inputs or outputs, an empty VHDL entity needs to be created as follows:

```
entity  testbench entity is
end entity testbench;
```

Next, the architecture of the testbench must contain an instantiation of the DUT. For that, one could instantiate the DUT using either component or direct entity instantiation. When using component instantiation, the component needs to be defined before using it in the main code. Also, a separate VHDL package could be used before the main code (in the same way as a signal).

The sample code below shows the syntax that is used to declare a component in VHDL. The component and port names must match the names used in the entity of the DUT.

```
Component and_gate is
    Port (
              A : in std_logic ;
              B : in std_logic ;
              Y : out std_logic
            ) ;
end component and_gate ;
```

Once component is declared, it needs to be connected as shown in the example below:

```
add_gate_instance : component and_gate
    port map (
          A => signal_A,
          B => signal_B,
          Y => signal_Y
                ) ;
```

It is important to note that every instantiation of the component must have a unique name. In the VHDL code example above, this is `and_gate_instance`.

The names on the left-hand side of the port map correspond to the names of the component ports. The names on the right-hand side are the signals which connect to the ports. These signals must be declared before using and must also have the same type of signal as the ports to which they connect.

The second technique is direct entity instantiation. When using this method, there is no need to declare a separate component. The example below shows the syntax for instantiating an AND gate using direct entity instantiation.

```
and_gate_instance : entity work.and_gate(rtl)
    port map (
        A => signal_A,
        B => signal_B,
        Y => signal_Y
                ) ;
```

As with the component instantiation technique, each direct entity instantiation must have a unique name as well. In addition, for direct entity instantiation, there are two additional requirements—the library and the architecture must also be specified. In the example above, the library is called work and an architecture called `rtl`.

One of the key differences between testbench code (VHDL) and design code (HDL) is there no need to synthesize the testbench. As a result, there are special constructs which apply timing to the testbench which is especially critical for creating test stimulus. There are two main constructs or statements in VHDL which can be used for timing. The most basic methods in the VHDL are the following:

- After (can either use concurrently or within process)
- Wait (this is only valid within process).

The VHDL code below shows the syntax that is used for the `after` and time statement:

```
<signal> <= <initial_value>,
                <end_value> after <time> ;
wait  for  <time> ;
```

In general, the `wait` statement in VHDL is to suspend execution of code for a given period of the time. The wait statement `wait until` is used to suspend the execution of code within a process until a given logical expression evaluates as true and is shown below:

```
wait  until <condition> for  <time> ;
```

However, `wait until` also uses the rising edge or falling edge macros to wait for specific events to occur. For example, using wait until statement for two signals (`sig_a and sig_b`) are set to logical one:

```
wait  until (
        sig_a  = '1'  and  sig_b = '1' ) for 2 us ;
```

In addition, `wait` statement or `wait  on` can be used to suspend execution of code until a signal changes state:

```
wait  on  sig_a, sig_b ;
```

Another consideration when writing a VHDL testbench is to generate a clock and reset signal. The `after` statement can be used to generate the signal concurrently in both conditions. For example, to generate a clock by scheduling an inversion every 1 ns, giving a clock frequency of 1 GHz that gives a fast simulation time and provides testbench clock frequency which should match the hardware clock as shown below:

```
-- Clock and Reset  (comment line)
clock <= not clock after 1 ns ;
reset <= '1', '0' after 5ns ;
```

The final part of the testbench is to code the test stimulus. For example, for a DUT requiring two inputs AND gate, followed by a sequential component, such as D-type Flip-Flop, one needs to generate each of the four possible input combinations. `Process` is used to generate this stimulus. In doing so, one needs to assign the inputs a value and then use a wait statement to allow for propagation through the DUT. The VHDL code example for testing two input AND gate driving a D-type Flip-Flop is shown below:

```
stimulus :
process begin
    wait until reset = 0 ;
    and_in <= "00" ;
      wait for 2ns;
    and_in <= "01" ;
      wait for 2ns;
    and_in <= "10" ;
      wait for 2ns;
    and_in <= "11" ;
      wait for 2ns;
wait
end process stimulus ;
```

The example below illustrates the discussed topics for testbench design using VHDL for a simple DUT and builds a testbench which generates every possible input combination. The DUT in this example consists of two input AND gate and its output connected to the D input of a D-type Flip-Flop.

```vhdl
entity testbench is
end entity testbench;

architecture test of testbench is
signal clock : std_logic := '0' ;
signal reset : std_logic := '1' ;
signal AND_input : std_logic_vector(1 down 0) :=
                              (others => '0') ;

Alias   A_input is AND_input(0) ;
Alias   B_input is AND_input(1) ;
signal Q_output : std_logic;

begin -- Define Reset and Clock Functions
  clock <= not clock after 1ns ;
  reset <= '1', '0' after 5ns ;

-- Instantiate the Design Under Test (DUT)
DUT: entity work.design(RTL)
port map (
  A => A_input,
  B => B_input,
  Q => Q_output) ;

-- Generate the test vectors or stimulus
stimulus:
process begin
-- Apply Reset
wait until (reset = '0') ;

-- Apply stimulus to Inputs for 2 clock periods apart
-- allowing propagation times.

   AND_input <= "00" ;
      wait for 2ns;
    AND_input <= "01" ;
      wait for 2ns;
    AND_input <= "10" ;
      wait for 2ns;
    AND_input<= "11" ;
      wait for 2ns;
wait
end process stimulus;
end architecture testbench;
```

1.4 Summary

In this chapter, the topic of ASIC design requirements is discussed. As mentioned, collecting proper ASIC design requirements is one of the most important steps for any given design project. The implementation process of designing an ASIC is considered a transformation of one stage of design to another one in sequential order, which takes a requirement from concept to an end-product (i.e., working silicon).

In addition, the concept of creating ASIC design architecture from requirements and producing design specification and initial design activities such as development using VHDL testbench is discussed.

Because the end-product is typically very complex and extremely small (in mm^2) in chip area, the design, implementation, and production processes are full of challenges and trade-offs which the system architects, designers, and product engineers need to make the best engineering decisions to meet design requirements and conformances.

References

1. *Business Analysis Body of Knowledge,* International Institute Business Analysis, April 15, 2015
2. Jayaram Bhasker, *Guide to VHDL Syntax: Based on the New IEEE Std 1076-1993,* Prentice Hall PTR, 1994

Chapter 2
Design Validation

*Design is concerned with how things work, how they are
controlled, and the nature of interaction between people and
technology. When done well, the results are brilliant,
pleasurable products.*

–Dan Norman

Validation is a pre-silicon and post-silicon process in which the ASIC is tested for
all functional correctness in a lab setup as well as using system level simulation. The
lab setup is done using Field-Programmable Gate Array (FPGA) assembled on a
test board, or commonly known as a reference board, along with all other component parts of the system for which the ASIC was designed. Often system-level simulation is considered the first step during validation and is to ensure correctness of the
ASIC design specification based on its requirements being coded using VHDL or
HDL (also known as RTL). The goal is to validate all use cases of the ASIC design
that a customer might eventually have in a true deployment and to qualify the design
for all these usage models prior to the ASIC continuation of the design implementation process or activities.

Validation starts initially for individual features and interfaces of the ASIC
design and then can also involve running real software/applications that stress tests
all the features of the design. A validation team usually consists of both hardware
and software engineers as the overall process involves validating an ASIC design in
a system-level environment with real software running on the hardware (i.e., reference board).

A typical ASIC design validation follows the below structure and can be broken
down into multiple steps. Some of these phases happen in parallel and some sequentially. In general, these steps are as follows:

- FPGA design
- FPGA programming
- Reference board design
- Hardware and software validation

© The Author(s), under exclusive license to Springer Nature Switzerland AG 2024

K. Golshan, *ASIC Design Implementation Process*,

https://doi.org/10.1007/978-3-031-58653-8_2

2.1 FPGA Design

Today, many ASIC designers are dependent on FPGAs to accelerate validation before software development and validate the ASIC design before committing to silicon.

A basic FPGA architecture contains a very large collection of unconnected digital components. This includes basic components such as multiplexers, logic gates, and Flip-Flops. An advanced FPGA architecture includes more complex components such as DSP core, memory blocks, and PLL core. The process of designing an FPGA is to connect these different components to create a complex system which is fundamentally designing hardware by creating an FPGA-based design or prototype. As a result, one can design a number of circuits which run in parallel to each other.

FPGAs can perform many different operations at the same time. This is a major advantage over the software approach, which require simulators be run extensively to cover different scenarios. In addition, there is much more control over the timing of design in FPGA. How long it will take to complete an operation can be easily estimated within a fraction of time. The same task could not be accomplished by using software simulation (i.e., hardware vs. software implementation).

As a result of these features, FPGA designs can be quicker than the equivalent implementation in a simulation environment. The drawback is that they tend to be more difficult to work with. This is not because prototyping FPGAs is inherently more difficult. The major difference is that there is a much smaller community of people who develop FPGAs as less libraries and open-source code are available. FPGA architecture consists of:

- Configurable Logic Blocks (CLB)
- Input/Output (I/O)
- A network of interconnection Switch Matrix (SM) between them
- Embedded SRAM memories
- Specialized components, such as DSP cores and PLL, in an advanced FPGA

The CLB blocks are the main building blocks of an FPGA. They are considered reconfigurable logic gates which perform different logical functions. Figure 2.1 shows basic FPGA physical configurations.

A typical CLB consists of a few inputs, LUT, multiplexer, and, perhaps, a RAM block. LUTs are simply small blocks of memory which are programmed to implement a given logical function. The exact structure of a CLB depends on the actual FPGA chip providers.

The I/O blocks provide connections to circuits which are external to the FPGA. They are connected directly to the physical pins of an FPGA and function as inputs, outputs, or mixture of both (bi-directional) connecting to select different standards of logic. In general, these I/O blocks support 3.3V (3V3) and/or 1.8 V (1V8). In addition, some advanced FPGA have dedicated pins for special functions, such as high-speed data transfers. The exact structure of an I/O block in an FPGA

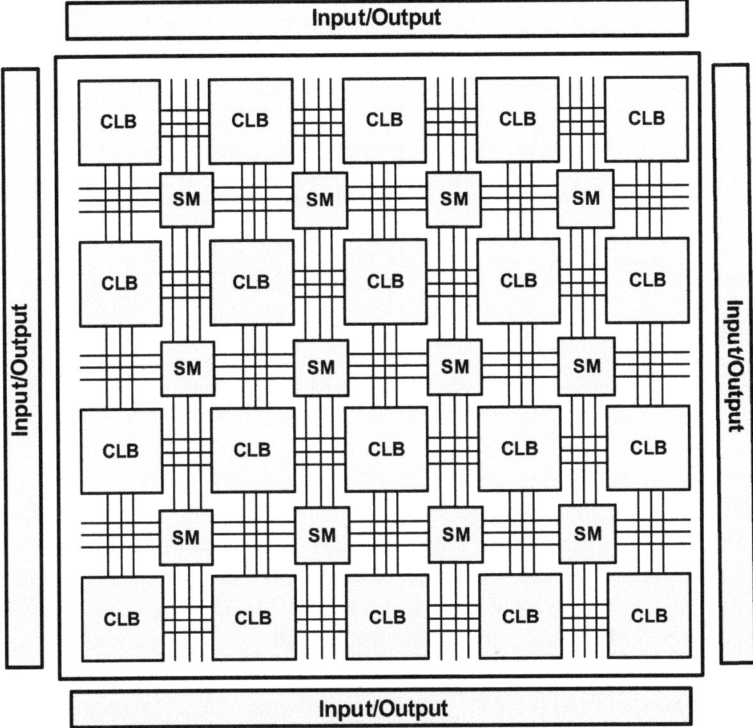

Fig. 2.1 An FPGA physical configuration example

varies slightly between different vendors. However, they typically include simple registers, such as D-type Flip-Flops and tri-state buffers.

Today's FPGAs have the capability to contain a complex and advanced ASIC design. However, in some cases when a large ASIC design is involved, there is a requirement for these designs to be partitioned among several FPGAs for prototyping [2] as shown in Fig. 2.2. This figure shows an ASIC design partitioned into four separate FPGAs (A, B, C, and D).

Splitting the design into several FPGAs can create various partitioning issues, especially for relatively large ASIC designs with complex connectivity. These issues can possibly be avoided if certain FPGA design guidelines, such as minimizing amount of interconnect between the FPGA, are followed.

As devices being prototyped on FPGAs are getting larger, following good design practices becomes more important for all ASIC design. Adhering to recommended synchronous design practices makes designs more robust and easier to debug. Using an incremental compilation flow adds additional steps and requirements but, however, can provide significant benefits in design productivity by preserving the performance of critical blocks and reducing compilation time.

As the industry becomes more competitive, time-to-market is one of the critical factors in an ASIC design process. To minimize the time-to-market, there is a need

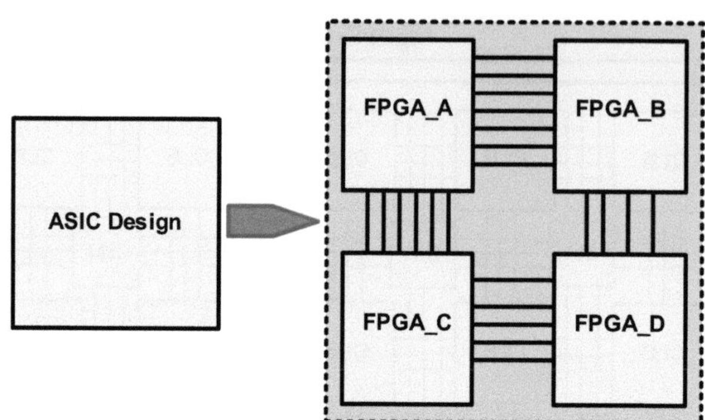

Fig. 2.2 An ASIC design partitioned into four FPGA examples

to ensure first time fully functional silicon avoiding re-spins, which can be achieved by extensive validation of design using FPGA prototyping.

The first step in FPGA design is to analyze whether the design will fit into a single FPGA. If the design is relatively small, it can be prototyped by just using a single FPGA. This reduces any need for design partitions or issues arising during or post partition. But for the modern intricate complex ASIC designs having large gate counts, a single FPGA might not be suitable to fit the whole of the design. So, there will be a need for partitioning the design into several FPGAs, and there will be partitioning issues that need to be taken care of during the design process.

For some ASIC designs, even though it might seem that the design will fit well in a single FPGA based on the ASIC design gate count, the design may still need to be partitioned because of the limited number of available CLB or resources, such as logical components on the targeted FPGA device, or the design is memory or DSP intensive (which is typical in ASIC designs). Once it is established that an ASIC design partition is required, there is a need to estimate the number of FPGAs required for the design prototype. As a part of the partitioning process, the following are major concerns which need to be addressed during partitioning an FPGA:

- The boundary of partition for a design is often guided by factors such as capacity of each FPGA and gate count of the partitioned design, availability of global clocks and number of asynchronous clock domains in the design, etc. Hence, one should keep them in mind while deciding upon the blocks to be partitioned.
- Understanding limitation on the number of available I/Os, interconnects (i.e., SM), and CLB in the FPGA, most ASIC designs may exceed the number of available I/O in the FPGA. One consideration is to either multiplex these ports before getting them to the connecting ports (pins) or changing the partition boundary in order to reduce the number of interconnecting nodes.
- As it is equally important to meet the timing of a partitioned design, the pin availability on FPGA is also constrained by timing factors, making it necessary to avoid poor pin assignment techniques.

- The presence of components such as memories, First-In-First-Out (FIFO)s, etc. which require mapping to FPGA resources such as RAM blocks, PLL, etc. on each FPGA need special attention in the partitioning process. They are limited by the architecture of FPGA and size of the device. This necessitates an intelligent demarcation to prevent exceeding the available resources as well as avoiding overutilization of them, as it might lead to routing and timing issues.

For prototyping FPGA, there are some suggested guidelines to avoid common partition issues such as:

Partition by design hierarchy   Actual ASIC design hierarchy must be considered during the partitioning process as it can simplify the process and eliminate the issue of re-partitioning the FPGA for minor changes (more so in cases where the ASIC design is not mature enough and minor changes happen more often). If a hierarchical instance is assigned as an FPGA partition, the partition should also include the entities instantiated below that instance which are not defined as separate partitions in order to eliminate cross partition interconnections.

Partition by design functionality   The FPGA design could be partitioned along functional boundaries, as in a top-level block diagram where each block is a design partition. Each block of a system can be considered as an independent module having more signal interaction internally than with other blocks. Keeping functional blocks together means that the FPGA synthesis tools can optimize related logic, which can lead to improved overall optimization. Also, the number of partitions required needs to be estimated first to determine the size of each partition as compiling time depends upon the partition size. Too many partitions should also be avoided as it can reduce the quality of results by limiting optimization.

Partition by clock domain and critical timing paths   At the early stage of an FPGA partitioning, clocks which feed the logic in each partition should be identified, and, if possible, clock domains should be kept within one partition.

When a clock signal is isolated to one partition, it reduces dependence on other partitions for timing optimization as the logic in the other partitions is no longer dependent on the clock signal present in the current partition.

Additionally, limiting the number of clocks within each partition simplifies the timing requirements for each partition during optimization. Logic which is implemented in design for clock domain transfers (such as a synchronization circuit) can be included inside the partition at one side of the domain transfer.

Isolating timing-critical logic from logic that is expected to meet its timing is critical as it preserves satisfactory results for non-critical partitions and focuses optimization iterations on just the timing-critical portions of the FPGA design, thereby minimizing compilation time. Also, there may be inter FPGA clocks (i.e., there is a clock that originates in one FPGA and needs to be used by the logic in the other FPGA) which may cause issues regarding timings and data coherency. In cases like these, it is advisable to route these clocks to the global clock pins located on interconnects. However, these clock pins should be constrained for input and output of the FPGA partition and aligned with respect to clocks.

Partitioning Inputs/Outputs by registers There are some input and output connections in design that are potentially timing-critical. Those connections can be partitioned using registers at an FPGA partition boundary, as boundary registers minimize the delays on inter-partition paths and prevent the need for cross-boundary logic optimizations. Partitioning this way gives routing delay, which needs to be considered for each register-to-register timing path, and hence, timing paths between partitions are likely not timing-critical, and each partition can be placed independent of other partitions. Additionally, the partition boundary does not affect combinational logic optimization because each register-to-register logic path is contained within a single partition. Registering every partition output ensures that the input timing performance for each design block is controlled exclusively within the destination logic block.

Minimizing across design partition and I/O boundary Minimizing the number of I/O paths crossing between partition boundaries makes partitions more independent for both logic and placement optimization. This is important for timing-critical and high-speed connections between partitions, especially in cases where the input and output of each partition are not registered.

In addition, during FPGA partitioning there are interconnections, among partitions which may need to be considered, such as:

- Connections that are not timing-critical are acceptable because they should not impact the overall timing performance of the FPGA partitioning. However, if there are timing-critical paths between partitions, inter-partition paths can be avoided by merging the partitions. The types of functions at the partition boundaries also govern the partitioning process. Partition quality of a design with merged partitions can be further improved, by adding registers to one or both sides of the cross-partition path subject to functionality adherence.
- If interconnects across partitions are in the same clock domain, partition boundary can also be simplified using an I/O multiplexing approach which involves multiplexing several interconnect signals on a single interconnect pin. The number of signals that can be multiplexed on a single interconnect pin depends upon the frequency requirements of the design as multiplexing X number of signals on a single pin reduces the maximum achievable frequency for the prototyped FPGA by approximately $1/X$ times.
- Another approach to minimize connections between partitions is to avoid using combinational such as glue-logic between partitions by moving the logic to the partition at one end of the connection in the other to keep multiple logic paths within one partition. For example, the connections between the partitions can be minimized by moving the glue-logic to the partition that has less connections, thus minimizing the interconnections between the two partitions.

2.2 FPGA Programming

Once the logical FPGA partitioning and design are completed, the FPGA need to be physically programmed. The programing of an FPGA is very much like an ASIC design flow with the exception that the end-product is an actual programmed FPGA device rather than a Geometrical Data Set (GDS) which needs to be fabricated by ASIC foundries. The steps that are required for a given FPGA design are as follows:

- Simulation
- Synthesis
- Placement
- Routing
- Timing analysis (statistic and/or gate-level simulation)
- FPGA programing file generation (Bitstream)

For simulation of an FPGA design, one needs to create a testbench similar to the one discussed in Chap. 1, to generate a number of inputs to the FPGA design. Then the FPGA design outputs need to be checked to ensure they are as expected (by either a manual inspection or via a self-checking). This process can be repeated using FPGA design functional code and post placement and routed net list. This is a model of the FPGA which was created by software tools when the FPGA design was implemented. This model includes information about the internal timing of the FPGA design and is more representative of the final implementation.

Typically, simulation is the main process that involves the verification of an FPGA design. This also could be complemented with hardware testing to ensure that the FPGA interfaces as expected with all external circuitries. However, as FPGA designs have become more complex, other techniques have become popular. More modern verification activities include Hardware-in-Loop (HiL) and emulation [1]. In both cases, this involves running code on an FPGA target device and feeding back data to simulation software. This allows the users to run specific, structured tests on their device in near real time.

Once FPGA RTL design code proves its correctness, one needs to program the FPGA design. Programming FPGA design requires synthesis, placement, routing, and programing file (Bitstream) generation.

The synthesis process is transforming the FPGA RTL into several interconnected gate levels of CLBs that are specific and internal under programing for a given FPGA. This includes mapping the netlist from synthesis to actual FPGA CBL resources through placement process.

Once FPGA design physical placement is completed, then routing among assigned CBLs through the routing process is required. It is normally necessary to run the place and route process several times to meet the timing requirements of an FPGA design using Static Timing Analysis (STA) and/or gate-level simulation. It is important to note that, if the FPGA design fails the STA or gate-level simulation, there is no guarantee that FPGA design will work reliably. When this occurs, then

Fig. 2.3 An example of an FPGA development board

either redo the implementation process with different settings, or change the FPGA design.

The final stage in the FPGA design implementation process is generating the programming file (Bitstream) which configures the FPGA. The Bitstream file is used to physically program an FPGA using an FPGA development board which is generated by the FPGA compiler. For programming an FPGA, there are various FPGA development boards available depending on FPGA manufacturers. Figure 2.3 shows an example of an FPGA development board.

2.3 Reference Board Design

Reference board design is a critical part of any ASIC design validation/development process to ensure the final end product ASIC operates based on its system architecture requirements with respect to its hardware design and associated operational software. The reference board validates that the ASIC design is functional in the system via hardware and software prior to fabrication

Reference board is used to validate all functional correctness of an ASIC design pre-silicon and post-silicon in the lab setup. This is done using the FPGAs (pre-silicon) and actual ASIC (post-silicon) assembled on a reference board along with all other components of the system for which the ASIC was designed. The goal is to validate all use cases of the ASIC that a customer might eventually have in a true deployment and to qualify the design for all these usage models. Reference board allows designers (hardware and software) to validate initially for individual features and interfaces of the ASIC design and can also involve running real software/applications that stress tests all the features of the ASIC design

To use reference board design for an ASIC design validation using FPGAs, one can design custom reference boards or purchase off-the-shelf FPGA-based reference boards. Often off-the-shelf boards don't satisfy the requirements for complex systems ASIC design. They're also expensive and lack scalability.

With the high cost of ASIC, one can hardly afford to make any mistakes in their ASIC design. The need for a good validation reference board is, hence, paramount. Hardware-based validation in-house reference board designs have been used ever since the advent of ASIC designs. More recently, FPGA-based references have started to address this need. However, off-the-shelf FPGA-based systems may not be ideal for the following reasons:

- *Cost:* If the reference board requires a limited number of FPGAs, the cost associated may not be justifiable. However, for a complex ASIC that often require multiple systems for different groups to use, the cost of off-the-shelf FPGA reference board design becomes prohibitively high.
- *Feasibility:* A system with a newer/critical prototype is difficult to create using off-the-shelf boards. Mapping the system design into off-the-shelf boards can be quite a challenge, especially if the system design has wide buses. Mapping a complex system design onto off-the-shelf boards can also reduce the frequency of operation significantly.
- *Performance:* Hardware and software engineers might like to design their own reference boards and validate an ASIC design out at or near the required frequency of its operation. Often the off-the-shelf boards have limited operating frequency. Off-the-shelf boards provide an expensive, partial solution that doesn't enable the users to validate their design and test it fully. Growing system design complexity, an increasing amount of embedded software, and the high cost of emulation systems have driven the need for custom reference board design and prototyping.

There is a significant complexity in designing reference boards, especially when multiple FPGAs are involved. Typically, as shown in Fig. 2.4, the process of designing a reference board is to use a synthesis tool to synthesize/partition the ASIC RTL and calculate the equivalent FPGAs resources, or CLB, using an FPGA compiler. Once the number of FPGAs required is identified and the number of connections between the different FPGAs is determined, the reference board design schematics can be created (using EDA schematic capturing tools) by incorporating the FPGAs along with other sub-systems to compose the entire system design. It should be noted that reference board design is an iterating process regarding the partitioning of the ASIC RTL. In other words, the reference board design schematic needs to be refined for the optimal number of FPGAs and their interfaces. Figure 2.5 shows an example of ASIC RTL validation board with an ASIC design based on requirements as discussed in Chap. 1.

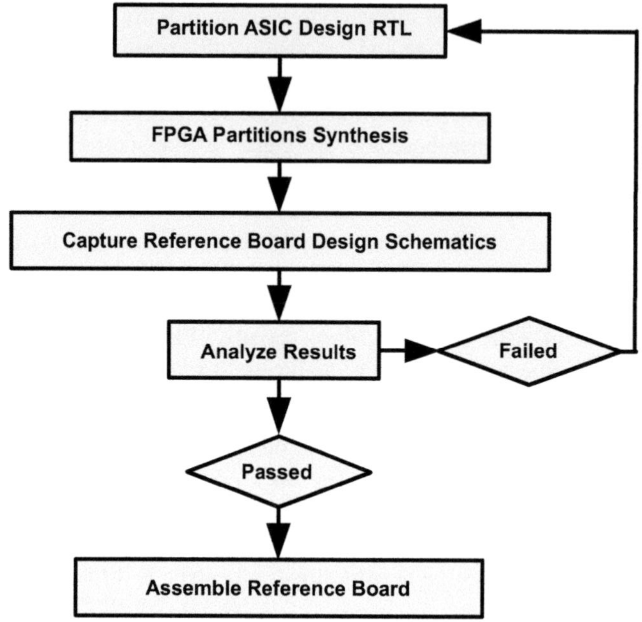

Fig. 2.4 Typical reference board design flow

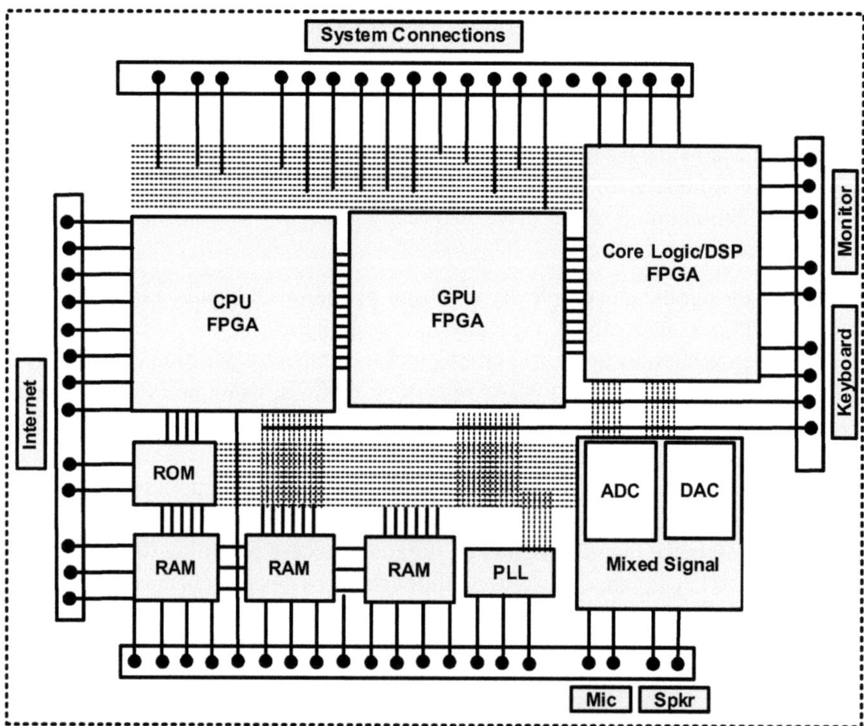

Fig. 2.5 An example of ASIC RTL validation board

2.4 Hardware and Software Validation

Hardware and software validation is based upon the process of embedded system design (i.e., reference board design) and generally starts with a set of requirements for what the product must do and ends with a working product that meets all the system requirements. System architecture defines the major blocks and functions of the system. Interfaces, bus structure, hardware functionality, and software functionality are determined. System designers use simulation tools, software models, and spreadsheets to determine the architecture that best meets the system requirements.

Once the architecture is set and the processor(s) have been selected, the next step is hardware design which includes component selection, Verilog/VHDL coding, synthesis, timing analysis, physical design of the ASIC, and the reference boards. The hardware design team will generate some important data for the software team such as the CPU, address map(s), and the register definitions for all software programmable registers. The accuracy of this information is crucial to the success of the entire project.

After the memory map is defined and the hardware registers are documented, the software design begins to develop many kinds of software such as boot code, to start up the CPU and initialize the system, hardware diagnostics, real-time operating system, device drivers, and application software. During this phase, tools for compilation and debugging are selected, and coding is done.

The most crucial step in reference board system design is the integration of hardware and software. Somewhere during the project, the newly coded software meets the newly designed hardware. How and when hardware and software will meet for the first time to resolve bugs should be decided early in the project. There are numerous ways to perform this integration. It's important to note that the integration needs to avoid debugging good software on broken hardware or debugging good hardware running broken software. This is because the concept of integrating hardware and software is system validation. These are the final steps to ensure that the working system meets the design requirements.

Although hardware/software co-validation has been around for many years, over the last few years, it has taken on increased importance and has become a validation technique used by more and more engineers. The trend toward greater system integration and demand for low-cost, high-volume consumer products has led to the evolution of the ASIC designs. An ASIC used to be defined as a single chip that includes one or more microprocessors, application-specific custom logic functions, and embedded system software. Microprocessors, and possibly DSPs, inside an ASIC have forced engineers to consider software as part of the ASIC design process to ensure correct operation. The techniques and methodologies of hardware and software co-validation allow projects to be completed in a shorter time and with greater confidence in the hardware and software.

Hardware and software co-validation is the process of verifying the embedded system software runs correctly on the hardware design before the ASIC design is committed for fabrication. Co-validation is often called virtual prototyping since the

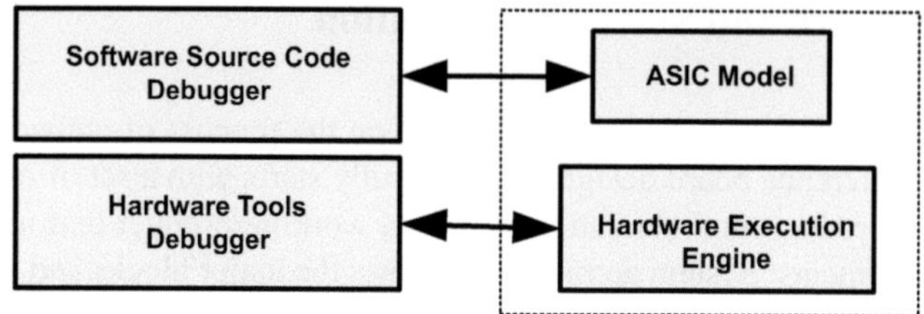

Fig. 2.6 Abstracted hardware and software co-validation example

simulation of the hardware design behaves like the real hardware but is often executed as a software program on a workstation. Figure 2.6 show abstracted hardware/software (HW/SW) co-validation for an ASIC design.

The ASIC model in Fig. 2.6 will be replaced by pre-silicon (i.e., FPGA design) and post-silicon (i.e., fabricated ASIC) in the reference board.

HW/SW co-validation provides two primary benefits. It allows software that is dependent on hardware to be tested and debugged before a reference board is available. It also provides an additional test stimulus for the hardware design. This additional stimulus is useful to augment testbenches developed by hardware engineers since it is the true stimulus that will occur in the final product of an ASIC. In most cases, both hardware and software teams benefit from co-validations. These co-validation benefits address the hardware and software integration problem and translate into a shorter project schedule, a lower cost project, and a higher quality product. The primary benefits of co-verification are:

- Early access to the hardware design for software engineers.
- Additional stimulus for the hardware engineers.

2.5 Summary

This chapter introduces an Application Specific Integrated Circuit (ASIC) design validation fundamental. Today, many ASIC designers are dependent on FPGAs to accelerate validation early in the process in order to start software development and validate the ASIC design before committing to silicon. A typical ASIC design validation follows the below structure and can be broken down into multiple steps. Some of these phases happen in parallel and some sequentially. In general, these steps are:

- FPGA design
- FPGA programming
- Reference board design
- Hardware and software validation

It explains the FPGA design, partitioning, and FPGA programing. In addition, the reference board design and the method of hardware and software co-validation are discussed.

References

1. Martin Schlager, *Hardware-in-loop Simulation*, KS Omniscriptum, April 2008
2. Michael John Sebastian Smith, *Application-Specific Integrated Circuits, FPGA Partitioning,* Addison Wesley Longman, January 1998

Chapter 3
Design Synthesis

Synthesis changed the very nature of how digital designs are
being constructed. –Aart de Geus

ASIC design synthesis is a crucial step in the process of creating custom integrated circuits tailored to specific applications. The synthesis process begins with the HDL description, such as VHDL or Verilog, which describes the intended functionality of the ASIC. The ASIC design may include modules, registers, combinational logic, and other elements. The goal of synthesis is to generate a gate-level netlist that represents the design in terms of logic gates, Flip-Flops, and other low-level components, including constant propagation, Boolean logic simplification, technology mapping, and resource sharing.

During synthesis, the HDL code is analyzed and transformed into a logical representation using various optimization techniques. These optimizations aim to improve the performance, power consumption, and area efficiency of the resulting ASIC. Some common optimizations involve the process of converting a high-level HDL design into a gate-level representation that can be implemented on a physical chip.

The synthesis process begins with a high-level HDL description, such as VHDL or Verilog, which describes the intended functionality of the ASIC. In general, the synthesis tool performs a top-down analysis of the design, starting with the highest-level modules and gradually expanding to the lower-level components. It identifies the required logic gates and Flip-Flops needed to implement the desired functionality and generates a netlist that specifies their interconnections.

Once the netlist is generated, it undergoes a technology mapping process, where the logical elements are mapped to specific standard cells from a target technology library. The library contains pre-designed standard cells, such as AND gates, OR gates, and Flip-Flops, which are available in the manufacturing process for the ASIC. The technology mapping process aims to select the most suitable standard cells from the library that can efficiently implement the desired logic.

© The Author(s), under exclusive license to Springer Nature Switzerland AG 2024
K. Golshan, *ASIC Design Implementation Process*,
https://doi.org/10.1007/978-3-031-58653-8_3

After technology mapping, the resulting gate-level netlist is further optimized to improve various aspects, such as area, power, and timing. Timing optimization ensures that the design meets the required timing constraints by adjusting the delays in the circuit. Power optimization techniques focus on reducing power consumption without compromising the functionality of the design. Area optimization aims to minimize the physical size of the ASIC by optimizing the placement and routing of the standard cells or gates.

Once the synthesis process is complete, the gate-level netlist is ready for physical design stages, which involve floor planning, placement, and routing to create the physical layout of the ASIC. The synthesized design can be further verified through simulation and formal verification techniques to ensure its correctness. The process of an ASIC design synthesis could be described as follows:

High-Level Design The process begins with an HDL description of the digital design, typically VHDL or Verilog. This description captures the functionality, behavioral, and constraints of the desire ASIC.

Design Analysis and Optimization The synthesis tool analyzes the high-level design and performs various optimizations to improve its quality. This may include logic minimization, constant propagation, resource sharing, and technology mapping. The goal is to reduce area, power consumption, and improve performance while preserving the desired functionality.

Logic Synthesis In this step, the high-level design is mapped to a gate-level representation using a library of standard cells or other predefined components. The synthesis tool selects appropriate gates and interconnects to implement the desired functionality. This gate-level netlist is a representation of the design using basic logic gates (AND, OR, NOT, etc.).

Technology Mapping The gate-level netlist is further optimized and mapped to specific cells from a target technology library, considering the available resources and constraints of the manufacturing process. This step involves choosing the best-fit cells for implementing the logic functions, considering factors like timing, area, power, and constraints specified by the target technology.

Timing Optimization Once the design is mapped to specific cells, timing analysis and optimization techniques are applied to ensure that the design meets the required timing constraints. This involves accounting for signal delays, interconnect delays, and other factors that impact the performance of the performance of the ASIC.

Design Verification After synthesis and optimization, the gate-level netlist needs to be verified to ensure that it behaves as intended. This typically involves functional simulation, formal verification, and other techniques to validate the correctness of the design.

Synthesizing an ASIC design involves various factors and considerations. While there isn't a definitive "best" way that applies universally, here are some key practices and guidelines that can help achieving a successful ASIC synthesis:

Clear Design Specification A well-defined and comprehensive design specification should outline the functional requirements, performance targets, power constraints, and any other specific requirements for the ASIC.

Choosing a Reliable and Efficient Synthesis Tool Select a reputable synthesis tool that suits the ASIC design requirements. Look for a tool that offers good support, optimization capabilities, and compatibility with the target technology library and, most importantly, their Quality of Result (QoR). The QoR is an important measure for ASIC designs that uses advanced node technologies.

Optimizing the ASIC Design's RTL for Synthesis Prioritize optimization techniques during the design process to improve the quality of synthesis results. This includes applying proper coding styles, using appropriate coding constructs, and leveraging synthesis-specific coding techniques like retiming, pipelining, and resource sharing.

Understanding and Utilizing the Target Technology Library One must gain a thorough understanding of the target technology library's cell offerings and characteristics. Proper utilization of the library, including understanding the available standard cells and their power and performance trade-offs and utilizing timing constraints, can significantly impact the synthesis results.

Performing Iterative Synthesis and Optimization Synthesis is an iterative process. After initial synthesis, analyze the synthesis report, and make iterative design changes, if necessary, to improve results. It's important to iterate design synthesis until the desired trade-off between area, power, and performance is achieved.

Timing Closure This is most critical for any ASIC design and requires careful attention to timing constraints in order to achieve timing closure. Timing analysis and optimization techniques, such as constraint optimization, retiming, and buffering, can help meet the required timing.

Validate Synthesis Results After synthesis, one needs to perform thorough verification to ensure the functionality and correctness of the synthesized design. This includes functional simulation, formal verification, and other verification techniques.

Collaborating with Physical Design Team It's helpful to work closely with the physical design team to ensure a smooth transition from synthesis to physical design. This collaboration must cover floor planning, placement, and routing to address physical design considerations and constraints.

As mentioned earlier, design synthesis is the process of translating the logical design into a gate-level netlist that can then be implemented as a physical silicon structure. The logical design and its detailed description are technology-independent until the synthesis process. The synthesis process uses advanced EDA tools that are aware of the capabilities and limitations of the target technology to which the high-level abstracted design is being ported. Design synthesis output is technology-dependent and tailored to the target ASIC fabrication or manufacturing process.

Once the RTL code and testbench are generated and validated, producing a gate-level description from RTL code is theoretically simple. The first pass gate-level design can typically be accomplished very quickly. Very few synthesis commands are needed to read RTL code, link the design, map to the targeted ASIC library, and produce the gate-level netlist. It should be noted that if the ASIC synthesis library (i.e., standard cell and other supporting components such as memories, etc.) has no errors, a first pass design will be logically correct. However, a synthesized design at this stage may not meet timing, size, power constraints, and Design for Test (DFT) of the ASIC design. The optimization of an ASIC for these parameters requires detailed design constraint specifications and can take many synthesis iterations. In general, the design synthesis flow is as follows:

- RTL synthesis
- Generate generic gate-level netlist using compiler built in its logical component library
- Mapping the generic gate-level netlist to the targeted standard cells library (first pass)
- Perform logic and timing optimizations based on the given constraints (numerous iterations)
- Generate targeted library netlist
- Convert sequential components to their equivalent scan (DFT)
- Generate the final place-and-route design netlist

Design synthesis requires model libraries from the ASIC fabrication vendor's technology. In general, these model libraries include component or cells such as standard cells, memories, and hard macros (e.g., PLL) which contain the following information:

- Descriptions (e.g., AND gate), rise/fall timing, intrinsic timing, pin information, input/output capacitance, signal types (such as input, output, and clock), logical functionality, and delay models
- Process parameters and area/power information
- Wire-load models, operating conditions, and scaling factors based on the operating conditions
- Attributes for various components
- Graphical symbols for netlist schematic generation that could be used for gate-level simulation

ASIC design specifications and validation design constraints are created and applied during design synthesis, place-and-route (i.e., physical design), as well as

STA. Each EDA tool attempts to meet these design constraints during the design's implementation process. These design constraints are categorized as:

- Timing constraints
- Optimization constraints
- Design rule constraints

ASIC designers use an industry standard format of Synopsys Design Constraints (SDC) to define timing, optimization, and design rule constraints. These constraints are essential to meet the ASIC design's goal (requirements and specifications) in terms of area, timing, and power to obtain the best possible implementation of the ASIC design.

The purpose of these design constraints is to ensure the design is functional and will behave correctly after manufacturing under various process-voltage-temperature (PVT) conditions.

3.1 Timing Constraints

The ASIC designer creates timing constraints for synthesis, physical design, and STA. These are a series of constraints applied to a given set of paths, or nets, that dictate the desired performance of a design. The major timing constraints are [1]:

- Clocks definition (period, frequency, or speed)
- Generated clock
- Virtual clock
- Clocks skew or uncertainty
- Multi-cycle path
- Case analysis
- False paths
- Input and output delays
- Minimum and maximum paths delay
- Disable timing arc
- On Chip Variation (OCV)

Synopsis Design Constraints (SDC) format is based on the *Tcl* format, and all commands follow the *Tcl* syntax. For timing constraints, the commands are related to timing specifications of the design which contain:

- Clock definition: `create_clock`
- Generated clock: `create_generated_clock`
- Virtual clock: `create_clock`
- Clock transition: `set_clock_transition`
- Clock uncertainty: `set_clock_uncertainty`
- Clock latency: `set_clock_latency`
- Propagated clock: `set_propagated_clock`

- Disable timing arc: `set_disable_timing`
- False path: `set_false_path`
- Input and output delay: `set_input_delay` and `set_output_delay`
- Minimum and maximum delay: `set_min_delay` and `set_max_delay`
- Multicycle path: `set_multicycle_path`

Clocks need to be defined with their source (i.e., port, pin, net, or virtual) and associated characteristics (i.e., period, duty cycles, skews, and rise and fall times).

Here is an example for defining clock source from output ports of a PLL (e.g., PPLO output port/pin).

```
// clock A 10ns with 50% duty cycle
create_clock -period 10 -name CIKA -waveform {0 5} [get_
ports PLLO]
```

Often, in an ASIC design, there is one or more internally generated clock sources such as clock dividers. In this case, one clock (CLKA) is external (chip input or a PLL output inside of the chip), and the other one (CLKB) is generated internally. As shown in Fig. 3.1, INST1 provides internal clock CLKB to INST2.

Assuming clock CLKA is created from PLLO port of PLL with a period of 10 ns and symmetrical waveform (50% duty cycle), then clock CLKB is considered the generated clock at Q port of INST1. The generated clock command is as follows:

```
// Generated clock A 10ns with 50% duty cycle
create_generated_clock -divide_by 2 -source CLKA -name CLKB \
 - waveform {3 13} [get_pins INST1/Q]
```

The corresponding waveforms between the master clock (CLKA) and its generated clock (CLKB) are shown in Fig. 3.2.

Another clock definition is virtual clock and can be defined as a clock without any source. In other words, a virtual clock is a clock that has been defined but has not been associated with any pin and port. The virtual clock does not physically exist in ASIC design. It is used as an external timing constraint for ASIC design interfaces by relating its inputs and outputs.

The virtual clock can be defined by the `create_clock` command without giving any generation point since there is no actual clock source in the design. There are non-existent launching and capturing registers that are used to time the design's input and output delay with a virtual clock.

The launch register represents the maximum external delay for design inputs for setup timing. The capture register represents the minimum delay for design outputs for hold timing. The design input and output delay with respect to virtual clocks is only valid after Clock Tree Synthesis (CTS). Figure 3.3 illustrates virtual clock concepts.

The commands to create a design's virtual clocks are as follows:

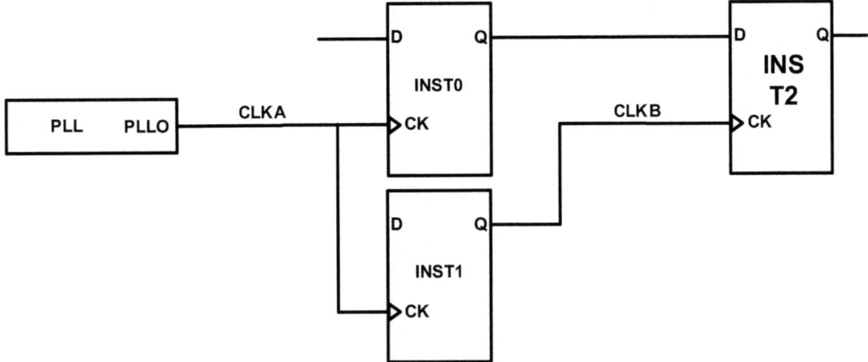

Fig. 3.1 Internally generated clock example

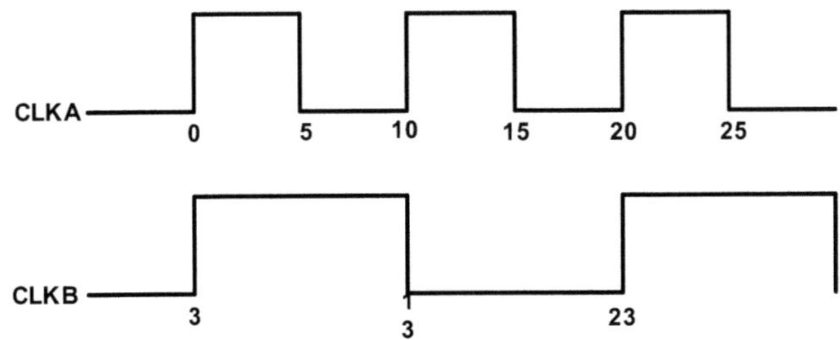

Fig. 3.2 Waveform presentation between CLKA and CLKB

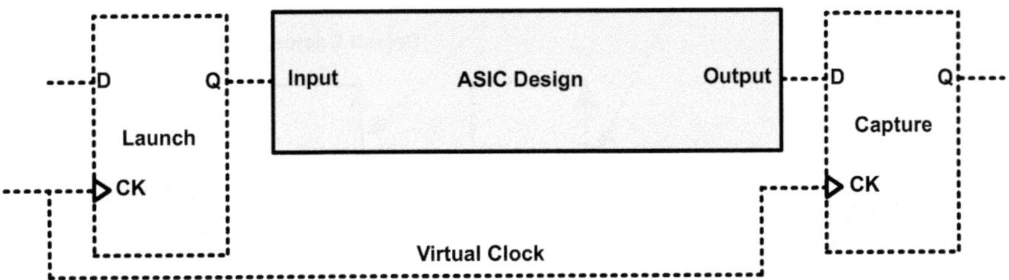

Fig. 3.3 Virtual clock concepts

```
//A virtual clock with 10ps period and 50% duty cycle
create_clock -name VIRTUAL_CLK -period 10 -waveform {0 5}

//Set input maximum delay of 4ns
set_input_delay -clock VIRTUAL_CLK-max 4 [get_ports Input Port]

//Set output minimum delay of 2ns
set_output_delay -clock VIRTUAL_CLK -min 2 [get_ports
Output Port]
```

Design synthesis tools consider the clock network delay to be ideal (i.e., a clock with fixed latency and zero skew) and are used during design synthesis. The physical design tools use the system clock definition to perform what is known as CTS and try to meet the clock networks' delay constraints [2].

Generally, the combinational data path between two Flip-Flops (launching Flip-Flop and capturing Flip-Flop) takes a single clock cycle to propagate the data through their associated combinational logic. For high-speed ASIC design, a single clock cycle is desirable. However, in some cases it could take more than a single clock cycle to propagate the data through the combinational data path. This is known as a multi-cycle path.

Using a multi-cycle path constraint, one could define the data as being captured by a captured Flip-Flop. The required capture edge would then only occur after the specified number of clock cycles. If a multi-cycle path is not defined, the setup check will occur after one clock cycle, and hold check will occur at the same edge of the captured Flip-Flop. This would cause a setup timing violation.

It should be noted that by default, hold check occurs at the edge of the clock prior to the capture or setup check as shown in Fig. 3.4.

Assuming the combinational data path is such that it takes three clock cycles to propagate data to the captured Flip-Flop, then there should be a multi-cycle path defined for the setup check between launch and captured Flip-Flops. Constraints for the multi-cycle path for setup and hold would be as follows (see also Fig. 3.5).

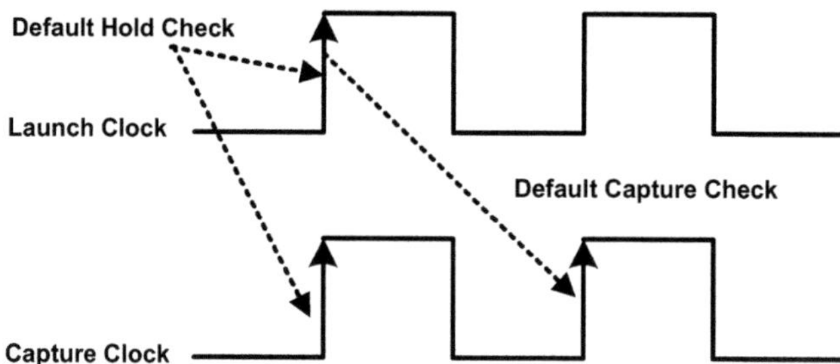

Fig. 3.4 Default hold and capture check

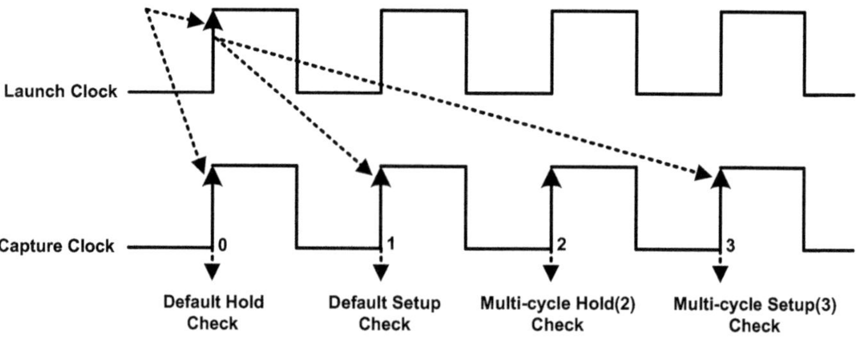

Fig. 3.5 Multi-cycle paths for setup and hold

```
// multi-cycle from setup check for launching to capturing clock
Set_multi_cycle_path -setup 3 -from [get_pins launching_flop/
output] \
                              -to [get_pins capturing_flop/input]

// multi-cycle from hold check for launching to capturing clock
Set_multi_cycle_path -hold 2 -from [get_pins launching_flop/
output] \
                              -to [get_pins capturing_flop/input]
```

During ASIC design synthesis, derating factors are used to account for process variation. They can be used for timing checks (setup and hold), net delay, and cell delay in the design. For example, if OCV concerns arise during the ASIC manufacturing process, derating factors can be used to address them.

A more realistic application of derating factors for advanced nodes is Advanced On Chip Variation (AOCV). In contrast to traditional OCV, AOCV derating factors increase as the distance increases and are, therefore, less pessimistic. AOCV is a distance-based (global) and path-based (local) model for derating [3].

3.2   Optimization Constraints

Optimization constraints are used to optimize speed, area, and power during synthesis and physical design. System clocks, and their delays and maximum area, are extremely important optimization constraints in ASIC design. System clocks are typically supplied externally, but they could be generated internally for a given ASIC design. All delays such as inputs and outputs, especially in a synchronous ASIC design, are dependent upon the system clocks. It is important to consider the relationship of an ASIC design's speed versus its area while setting these design optimization constraints. These constraints are applied by a means of cost functions. These cost functions are:

- Maximum timing delay cost
- Minimum timing delay cost
- Minimum power cost
- Maximum area cost
- Minimum area cost

Figure 3.6a, b show the relationships between an area and operational speed for a given ASIC design with respect to timing and area cost functions. As shown in these graphs, higher operational speed or frequency cost function results in larger device area (it should be noted that the maximum delay cost function has the highest priority in the cost function calculation during synthesis).

Superimposing Fig. 3.6a, b (as shown in Fig. 3.7) suggests a balance between maximum delay and maximum area cost functions as an optimal cost function

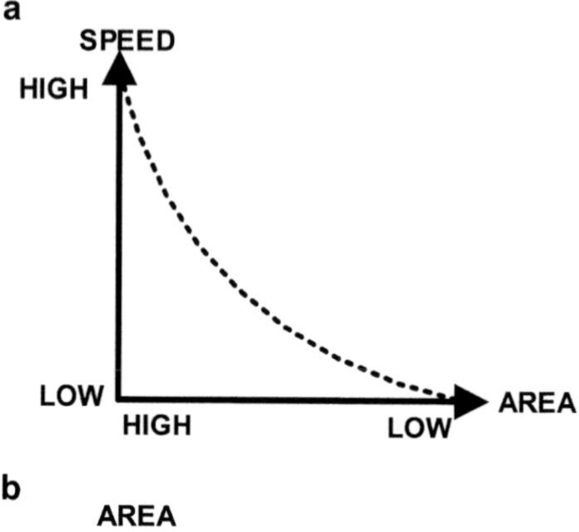

Fig. 3.6 (**a**) Speed vs. area. (**b**) Area vs. speed

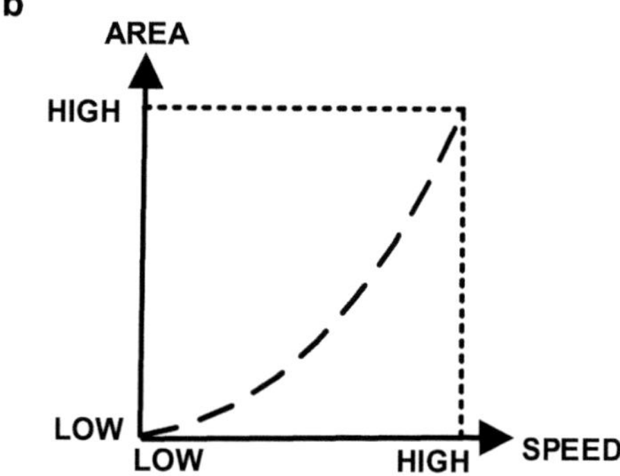

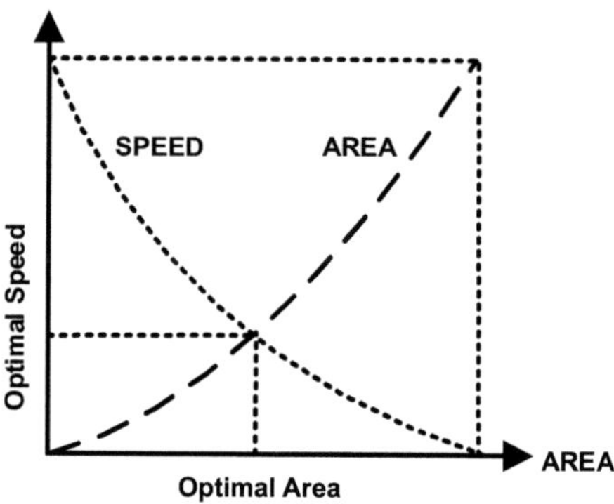

Fig. 3.7 Speed vs. area cost functions

setting. This is like defining a product price based on its supply and demand curve in economic terms. This means it's almost impossible to have a high-speed design with low silicon area.

3.3 Design Rule Constraints

Design rule constraints are used to setup the environment under analysis and physical implementation of ASIC designs. The most basic design rule constraints are:

- `set_driving_cell`
- `set_input_transition`
- `set_load`
- `set_max_fanout`
- `set_case_analysis`

If multi-voltage islands are present in the ASIC design, their corresponding design rule constraints are:

- `create_voltage_area`
- `set_level_shifter_strategy`
- `set_level_shifter_threshold`
- `set_max_dynamic_power`
- `set_max_leakage_power`

Input and output delays are used to constrain the boundary of external paths in an ASIC design. These constraints specify point-to-point delays from external inputs to the first registers and from registers to the outputs.

Minimum and maximum path delays provide greater flexibility for physical design tools that have a point-to-point optimization capability. This means that one can specify timing constraints from one specific point (i.e., pin or port) in the ASIC design to another, provided such a path exists between the two specified points.

Input transition and output capacitance loads are used to constrain the input slew rate and output capacitance of an ASIC device input and output pins. These constraints have a direct effect on the final ASIC design timing.

The values of these constraints are set to zero during physical design and place-and-route activity to ensure that the actual ASIC design timing is calculated independent of external conditions and to make sure register-to-register timing is met. Once that is achieved, these external conditions can be applied to the design for input and output timing optimization.

False paths are used to specify point-to-point non-critical timing, either internal or external, to an ASIC design. Properly identifying these non-critical timing paths has a considerable impact on physical design tools' performance. Design rule constraints are imposed upon ASIC design by requirements specified in each standard cell library or within physical design tools.

Design rule constraints have precedence over timing constraints because they must be met to realize a functional ASIC design. There are four types of major design rule constraints:

- Maximum number of fan-outs
- Maximum transitions

- Maximum capacitance
- Maximum wire length

Maximum number of fan-outs specify the number of destinations that one cell can connect to for each standard cell in the library. This constraint can also be applied at the ASIC design level during the physical synthesis to control the number of connections one cell can make.

Maximum transitions constraint is the maximum allowable input transitions for each individual cell in the standard cell library. This constraint can also be applied to a specific net or to an entire ASIC design.

Maximum capacitance constraint behaves similarly to maximum transitions constraint, but the cost function is based on the total capacitance that a standard cell can drive for any interconnection in the ASIC design. It should be noted that this constraint is fully independent of maximum transitions and, therefore, can be used in conjunction with maximum transition.

Maximum wire length constraint is useful for controlling the length of wire to reduce the possibility of two parallel long wires of the same type. Parallel long wires of the same type may have a negative impact on the noise injection and could cause crosstalk.

These design rule constraints are mainly achieved by properly inserting buffers at various stages of physical design. Thus, it is imperative to control the buffering during place-and-route to minimize area impact.

3.4 Summary

This chapter provides an overview of an ASIC design synthesis that transforms a high-level HDL description into a gate-level netlist, optimizing the design for performance, power, and area efficiency. It lays the foundation for the subsequent physical design stages and plays a vital role in the successful creation of application-specific integrated circuits.

In addition, an overview of design constraints that are used during ASIC design synthesis is provided. These constraints are categorized by their functions during synthesis. The types of constraints that outlined are timing, optimization, and design rules.

Timing constraints cover clock definitions such as system clocks, internally generated clocks, virtual clocks, multi-cycle paths, and false paths.

Optimization constraints cover speed, area, and power during synthesis and physical design.

Design rule constraints are imposed upon ASIC design by requirements specified in each standard cell library or within physical design tools.

References

1. Pran Kurup and Taher Abbasi, *Logic Synthesis Using Synopsys*, Kluwer Academic Publishers, 1995
2. Khosrow Golshan, *Physical Design Essentials, an ASIC Design Implementation Perspective*, Springer Business Media, 2007
3. Khosrow Golshan, *The Art of Timing Closure, Advanced ASIC Design Implementation*, Springer Nature, Switzerland AG, 2020

Chapter 4
Physical Design

ASIC physical design refers to the process of transforming a conceptual design of an IC into a physical layout that can be fabricated. It involves placement and routing of various components on a semiconductor substrate to meet the desired performance, power, and area requirements. The steps used in ASIC physical design consider various performance, power, and manufacturing constraints. The goal is to achieve an optimized layout that meets the design specifications and can be fabricated successfully.

Physical design phases may be viewed as transformations of the representation in various steps. In each step, a new representation of an ASIC design is created and analyzed. These physical design steps are iteratively improved to meet ASIC design requirements. For example, the placement or routing steps are iteratively improved to meet the design timing specifications.

Another challenge commonly faced by physical designers is the occurrence of timing and/or design rule violations during final ASIC design timing analysis and/or verification (logical and physical). If such violations are detected, the physical design steps need to be repeated to correct the errors. These error corrections have a direct impact on the ASIC cycle time because it may require the entire physical design steps be repeated to meet timing specifications and/or to correct verification errors [1]. The physical design activities typically consist of the following:

- *Floorplanning:* This stage involves dividing the chip area into different functional blocks and determining their approximate locations. It also includes defining the overall chip size, pin locations, and power and clock distribution networks.
- *Placement:* In this stage, the synthesized gate-level netlist is mapped onto the physical chip area. The goal is to find an optimal arrangement of the cells to minimize wire length, power consumption, and other metrics. Placement tools

© The Author(s), under exclusive license to Springer Nature Switzerland AG 2024

K. Golshan, *ASIC Design Implementation Process*,

https://doi.org/10.1007/978-3-031-58653-8_4

use algorithms to determine the location of each cell while considering various constraints.

- *Clock Tree Synthesis (CTS):* Clock signals are critical in ensuring synchronous operation within the circuit. CTS involves creating a tree-like structure that distributes the clock signal uniformly and efficiently across the chip, reducing skew and ensuring proper timing.
- *Routing:* Once the placement is complete, the routing phase begins. It involves determining the paths of metal interconnects (wires) to connect the various components of the circuit. Routing algorithms consider signal integrity, timing, and congestion while optimizing the interconnect density. During the routing stage, the design undergoes physical verification checks. These checks ensure that the layout adheres to design rules; avoids violations, such as shorts and opens; and meets the manufacturing requirements. This step helps identify and correct any potential issues before fabrication.
- *Design for Manufacturing (DFM):* DFM techniques are applied to enhance the manufacturability and yield of the chip. This includes measures like adding dummy fill and spare cells to ensure uniformity during the fabrication process, optimizing the design for lithography and mask generation, and considering other process-related constraints.

4.1 Floorplanning

In physical design, the floorplanning stage refers to the process of partitioning the chip area and determining the approximate placement of various components and modules of an IC. It involves making high-level decisions about the chip's organization, dimensions, pin locations, power distribution, and other macro-level considerations.

Floorplanning is the art of any physical design. A well-thought-out floorplan leads to an ASIC design with higher performance, optimum area, and low power consumption. Floorplanning can be challenging in that it deals with the placement of I/O pads and macros as well as clock distribution networks for extreme high-frequency clock power and ground structures.

Since the first step in any physical design is floorplanning, it is imperative to use high-quality and well-prepared data such as libraries (logical and physical), technology files (i.e., Library Exchange Format or LEF), design constraints, incoming pre-layout netlists (i.e., timing), and operational scripts. This becomes more important when dealing with advanced nodes (e.g., 40 nm and below) and usage of Multi-Mode Multi-Corner (MMMC). Therefore, a check and balance must be performed during the floorplan stage.

Before one proceeds with the physical floorplan, it's important to make sure that the data used during physical design activity is prepared properly. Proper data preparation is essential, not only to the ASIC design floorplanning but to all other stages of ASIC physical design in order to implement a correct-by-construction design. This is especially important when dealing with MMMC.

Assuming a given process node all logical and physical libraries have gone through an extensive QC flow, the first step for floorplanning is to modify configurational, general settings and floorplan scripts of the physical design tools to reflect project requirements.

Efficient design implementation of any ASIC requires an appropriate style or planning approach that enhances the implementation cycle time and allows the design goals, such as area and performance, to be met [2]. There are two style alternatives for design floorplanning—flat or hierarchical.

For small to medium ASICs, flattening the design is well suited; for very large and/or concurrent ASIC designs, partitioning the design into sub-modules, or hierarchical style, is preferred.

A flat floorplanning provides better area usage, but requires effort during physical design and timing closure compared to hierarchical floorplanning. This area advantage is mainly due to there being no need to reserve extra space around each sub-design partition for power, ground, and resources for the routing.

Timing analysis efficiencies arise from the fact that the entire design can be analyzed at once rather than analyzing each sub-circuit separately and then analyzing the assembled design later. The disadvantage of this method is that it requires a large memory space for data and run time increases rapidly with design size.

The hierarchical floorplanning style is mostly used for very large and/or concurrent ASIC designs where there is a need for a substantial amount of computing capability for data processing. In addition, it is used when sub-circuits are designed individually. However, hierarchical design floorplanning may degrade the performance of the final ASIC. This performance degradation is mainly because the components forming the critical path may reside in different partitions within the design, thereby extending the length of the critical path. To avoid performance degradation, it is possible to assign the critical components to the same partition or generate proper timing constraints to keep the critical timing components close to each other, thereby minimizing the length of the critical path within the ASIC design. In the hierarchical design floorplan, an ASIC design can be partitioned logically or physically.

Logical partitioning takes place in the early stages of ASIC design (i.e., RTL coding). The design is partitioned according to its logical functions, as well as physical constraints, such as interconnectivity to other partitions or sub-circuits within the design. In logical partitioning, each partition is placed and routed separately and placed as a macro, or block, at the ASIC top level.

Physical partitioning is performed during the physical design activity. Once the entire ASIC design is imported into physical design tools, partitions can be created. Most often, these partitions are formed by recursively partitioning a rectangular area containing the design and using vertical or horizontal cut lines. Physical partitioning is used for minimizing delay and satisfying timing and other design requirements in a small number of sub-circuits. Minimizing delay is subject to the constraints applied to the cluster for managing circuit complexity.

Initially, these partitions have undefined dimensions and fixed area (i.e., the total area of cells or instance added to the partition) with their associated ports, or terminals, assigned to their boundaries such that the connectivity among them is minimized. To place these partitions, or blocks, at the chip level, their dimensions as well as their port placement must be defined.

Perfection often requires iteration. It is recommended that once RTL design is nearing its design completion (e.g., 80% completed), the physical designer should complete the full physical design from floorplan through physical verification, power analysis (such a leakage), and STA to ensure readiness of the entire ASIC design implementation flow.

MMMC view definitions are used for concurrent timing analysis and are physical-design-stage dependent. During the floorplan stage, it is used for timing checks for all modes and corners in the design; during the placement stage, for setup timing analysis; during the CTS stage, for setup and hold timing analysis; and during the design's final route stage, timing analysis for all modes and corners.

Power and ground are primarily used during physical design floorplanning providing standard cells' power and ground mesh for core and macros power and ground. For technology nodes such as 90 nm and higher, the standard practice was to put power and ground next to each other. However, for advanced nodes, this type of placement will short the power and ground. This is due to advanced nodes having smaller metal layer spacing which causes the short between power and ground if there is particle contamination during the silicon process. To avoid this, one needs to alternate placement of power and ground.

Before starting floorplanning, the design needs to be timed by activating the MMMC mode for all functional modes for setup timing checks. At this stage hold timing checks are not required because all clocks are ideal. For setup timing checks, the design must meet required timing with no or very small violations due to differences between synthesis and physical design tools of their timing engines (i.e., timer).

If large setup timing violations are observed (e.g., larger than 70 ps or so), they need to be investigated before continuing. Large setup timing violations may have several sources. Design constraint issues (e.g., missing false paths, multi-cycle paths, and/or unrealistic external inputs/outputs delay) may be the cause of these violations. Or the design may have been synthesized using single mode rather than multi-mode. In this case, one could perform netlist-to-netlist optimization within the physical design tool option in order to optimize the incoming netlist using MMMC. This, of course, is dependent upon the tool supporting such an option. Otherwise, the MMMC method would need to be used during synthesis.

Once no major violations are found, one can proceed with floorplanning.

The first step would be to make the chip power and ground ring. Next, place hard macros, such as memories and PLL. Do not apply any power and ground strips and/or mesh. Instead run a coarse placement. After coarse placement, all modules in the design should be highlighted using different colors to see which modules are clustered together. If the modules are not clustered together, it is mainly due to the placement of hard macros. They need to be placed so that module clustering occurs. Proper module clustering improves timing and minimizes routing. However, if some of the modules are tightly clustered (i.e., area congestion too high), one could use the area utilization option of the design tool to reduce the congestion. High congestion of the area makes the region unrouteable and could adversely impact timing.

Floorplan refining is an iterative process until the optimal floorplan is achieved. Figure 4.1 illustrates a refined floorplan after coarse placement.

Often the checks that detect problems with the physical database are related to the netlist, such as unconnected ports, mismatched ports, standard cell errors, or errors in the library and technology files. The physical design tool generates a log file that contains all errors and warnings. It is important to review the log file and make sure that all reported errors and warnings are resolved before proceeding to the next phase.

With smaller process geometry (e.g., 20 nm or below), increasing speed and number of gates in ASIC design will increase power consumption, and that could be limiting for many applications. Hence, an important aspect of today's ASIC design is to manage power and reduce its consumption. There are various techniques for reducing the power [3] during the floorplanning stage. These techniques are:

- Multiple power supply and level shifter (low and high) insertion
- Power gating (shutting off a portion of the design)

Multiple power supply domains are used for conserving the dynamic and static power of the design. Different functional domains run at different supply voltages. In this way one can save the power losses by reducing the supply voltage for standard cells and memory elements in the design.

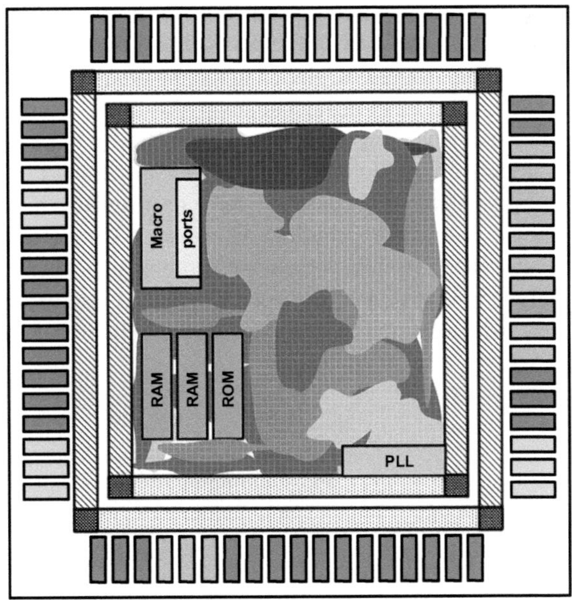

Fig. 4.1 Refined floorplan illustration

Different power domains are defined based on the criticality of the design. In this style of floorplanning, level shifters are used for the signal coming from the low-voltage power domain to the high-voltage power domain and vice versa. At the netlist level, the design code will be written in Cadence Power Format (CPF) or Universal Power Format (UPF) based on which one can develop the power structure for the design. Below is an example of level shifter insertion in CPF power format:

```
### High-to-Low level shifters ###
define_level_shifter_cell -cells LVS-H2L* \
-input_voltage_range 0.9:1.0:1.1 \
-output_voltage_range 0.8:1.0:0.1 \
-direction down \
-output_power_pin VDD \
-ground VSS \
-valid_location to

### Low-to-High Level Shifters ###
define_level_shifter_cell -cells LVS-L2H* \
-input_voltage_range 0.8:1.0:0.1 \
-output_voltage_range 0.9:1.0:1.1 \
-input_power_pin VDD-IOW\
-output_power_pin VDD \ U
-direction up \
-ground VSS \
-valid_location to
```

From a floorplanning perspective, there are some special placement guidelines for inserting the level shifter across the different power domains in the design. The level shifter should be placed in the destination domain of the design. There is one disadvantage of inserting the level shifter—it occupies area in design. But at the same time, inserting the level shifter will help in saving the power of the device.

Figure 4.2 shows high-to-low level shifter and Fig. 4.3 low-to-high level shifter.

From a timing perspective, these level shifters have a minimal impact on timing (they are similar to a buffer). However, it's important to note that for high-to-low level shifters, the low voltage swing input signal is not necessarily strong enough to fully turn on the input transistor. This could lead to an unacceptably long rise or fall time, which could then result in a higher switching current and reduced noise margin.

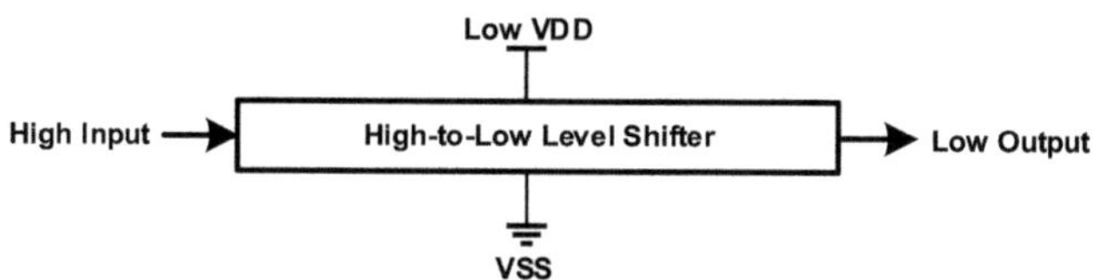

Fig. 4.2 High-to-low level shifter

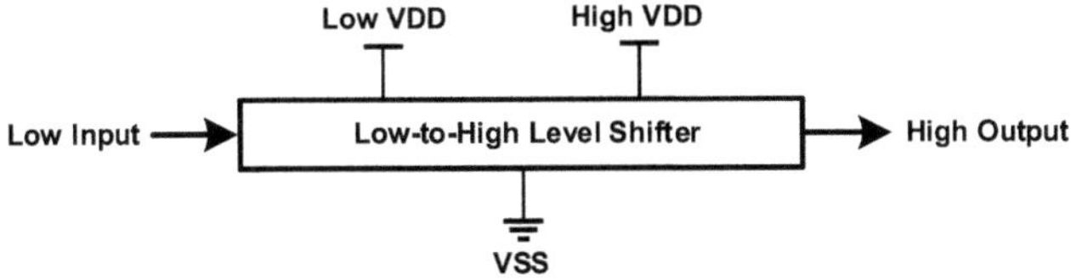

Fig. 4.3 Low-to-high level shifter

The last technique used for reducing the power is known as power gating. This is especially effective for leakage power reduction versus dynamic power reduction. In this technique, one would shut down a portion of the design by means of disconnecting their supplies either from the power or ground.

There are two types of power gating. One is fine grain which adds power-down transistors to every standard cell used in the shutdown domain. This can create timing issues introduced by inter-cluster voltage variations that are difficult to resolve.

The second type of power gating is a coarse grain which implements the grid style of power-down domain and drives standard cells through a virtually powered network. This approach is less sensitive to the process variations and introduces less power leakage. In coarse-grain power gating, the power-gating transistors are a part of the power distribution network rather than the standard cell. The blocks are placed in the shutdown mode when the function is not active and turned on when required, using two types of power switches. One uses PMOS transistors as the power-down cell. This is known as the header cell for the power supply (VDD) control.

The other method uses NMOS transistors as the ground cell. This controls the ground supply (VSS) and is known as the footer cell. Because of the effectiveness of using the header cell over the footer cell for power gating, the header cell will be discussed.

The header cell contains two sizes of PMOS transistors: one small and one large, regarding their gate length. Using a small PMOS transistor brings up the power-down domain (switched supply) slowly before the large PMOS transistor is turned on. This is done to prevent a large amount of current surge which has the potential of damaging standard cells' gates within the power-down domain during the powering up procedure.

Figure 4.4 shows a header cell circuit. In this example the header cell has one small PMOS transistor with input (IN1) and inverted output (OUT1). The same goes for the large PMOS transistor with input (IN2) and inverted output (OUT2).

From a physical design point of view, when the header cells are placed next to each other, they form two chains—a chain of small PMOS transistors and a chain of large PMOS transistors. The last small PMOS transistor in the chain will turn on the first large PMOS transistor in its chain.

The header cells are used to build a ring around the switched power domain.

Either CPF or UPF is used to define the header cell, isolation cells, and always-on and power-down domain. An example of the power-down domain's CPF is shown below:

Fig. 4.4 Example of
header cell

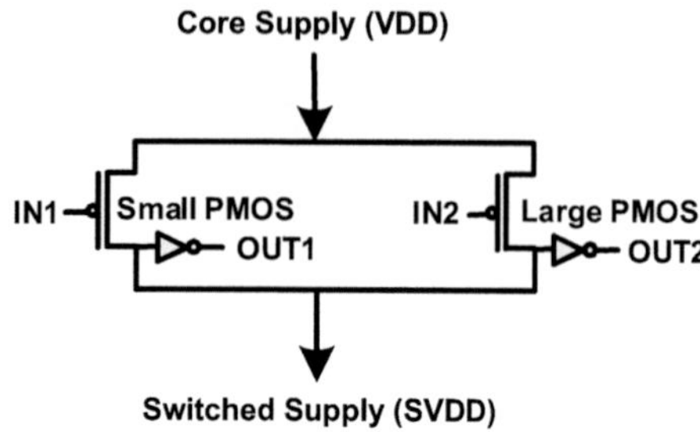


```
set_cpf_version 1.0
#### Isolation Cells ###
define_isolation_cell -cells ISO_AND \
-power VDD \
-ground VSS \
-enable ISO \
-valid_location to

### Isolation Rule ###
create_isolation_rule -name ISORULE -from PWRDOWN \
-isolation_condition "!PWRDOWN/isolation_enable" \
-isolation_output high

update_isolation_rules -names ISORULE -location to -cells ISO_AND


### Power Switch cell (header) ###
define_power_switch_cell -cells {HEADER} \
-power_switchable SVDD -power VDD \
-stage_1_enable !IPWRON1 \
-stage_1_output IPWRON2 \
-stage_2_enable !PWRON2 \
-stage_2_output I ACKNOWLEDGE \
-type header

### Always-On (AO) Domain ###
create_power_domain -name AO -default
create_power_nets -nets VDD  -voltage 0.8
create_ground_nets -nets VSS
update_power_domain -name AO -internal_power_net VDD
create_global_connection -domain AO -net VDD -pins VDD
create_global_connection -domain AO -net VSS -pins VSS

### Power Down (PWRDOWN) Domain ###
create_power_domain -name core -instances PWRDWN \
```

```
-shutoff_condition {PWRON1/pwron_enable}
create_power_nets -nets SVDD -internal -voltage 0.8
create_ground_nets -nets VSS
update_power_domain -PWRDOWN -internal_power_net SVDD
create_global_connection -domain PWRDOWN -net VSS -pins VSS
create_global_connection -domain PWRDOWN -net VDD_SW -pins SVDD
create_power_switch_rule -name PWRSW -domain PWRDOWN \
-external_power_net VDD
update_power_switch_rule -name PWRSW \
-cells HEADER \
-prefix PWR_SW_ \
-acknowledge_receiver ACKNOWLEDGE
```

During floorplanning, once the power-down domain region is defined, a header cell ring-chain is constructed around it. Figure 4.5 shows the structure of the header cell's ring-chain. Additionally, there are two cells—Start and End.

With the Start cell, when PWRON1 is low, the power domain is in the shutdown mode. The End cell provides a non-inverted signal from the last header cell's output. This is used as ACKNOWLEDGE (high) indicating the power domain is ON.

According to Fig. 4.5, in order to bring the power-down domain to an ON state, PWRON1 input signal needs to go from a low-to-high state which will turn on the small PMOS transistor chain.

Once all small PMOS transistors are turned on, its last stage triggers the first stage of large PMOS transistor chain by the PWRON2 input going from a low-to-high state. Once the large PMOS transistors are turned on, the power-down domain's power supply will be ON by connecting VDD (core supply) to SVDD (switched VDD) through the PMOS transistors' R-on resistance in the header cells.

To ensure their floating outputs of the power-down domain are not propagating into the always-on domain when the power-down domain is in shut-down mode, isolation cells are used. In this method, isolation cells are ANDed and/or NANDed with ACKNOWLEDGE with the outputs of the power-down domain preventing any floating output signals.

One needs to consider the IR drop across the header ring. Depending upon the header cell's layout, its R-on resistance could be obtained through Spice simulation. Once the header cell's R-on resistance is determined, it can be used to calculate the

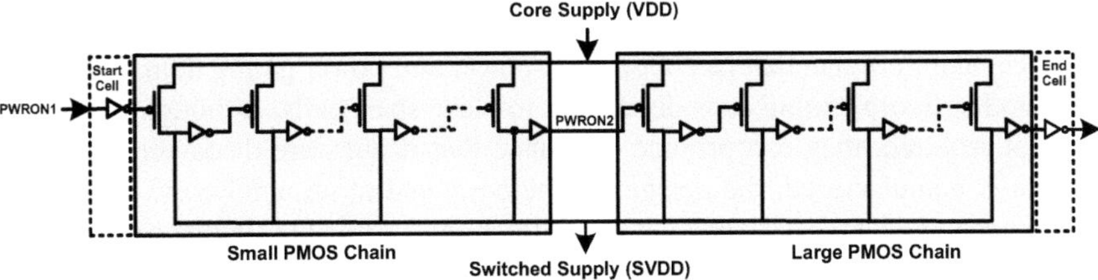

Fig. 4.5 Header cell ring-chain structure example

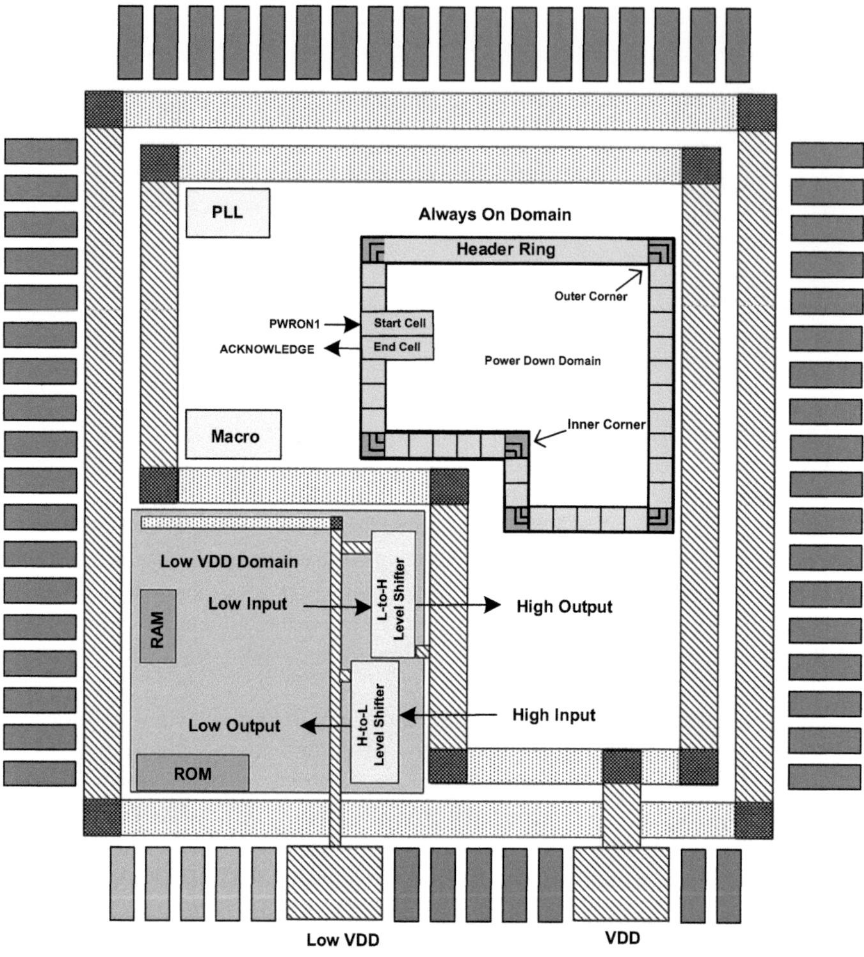

Fig. 4.6 Conceptual floorplan with different power domains

number of header cells in the ring. The higher the header cell count, the lower the ring IR drops.

In contrast to the chip's input/output pad ring, the header ring library contains header cells, header filler cells, start cell, end cell, power protection (commonly known as CLAMP) cells, outer corner cells, and inner corner cells. Figure 4.6 show a conceptual floorplan that has both multi-power and power gating domain styles.

Another floorplanning consideration is to place spare cells. Although spare cells are not required, they can provide insurance that if bugs are discovered after the design is manufactured, the design can be corrected at minimal cost. Spare cells allow one to go back later and do a small functional correction Engineering Change Order (ECO) to the design. The spare cell could be used from standard cell libraries and/or from special libraries called metal-programmable cell ECO libraries.

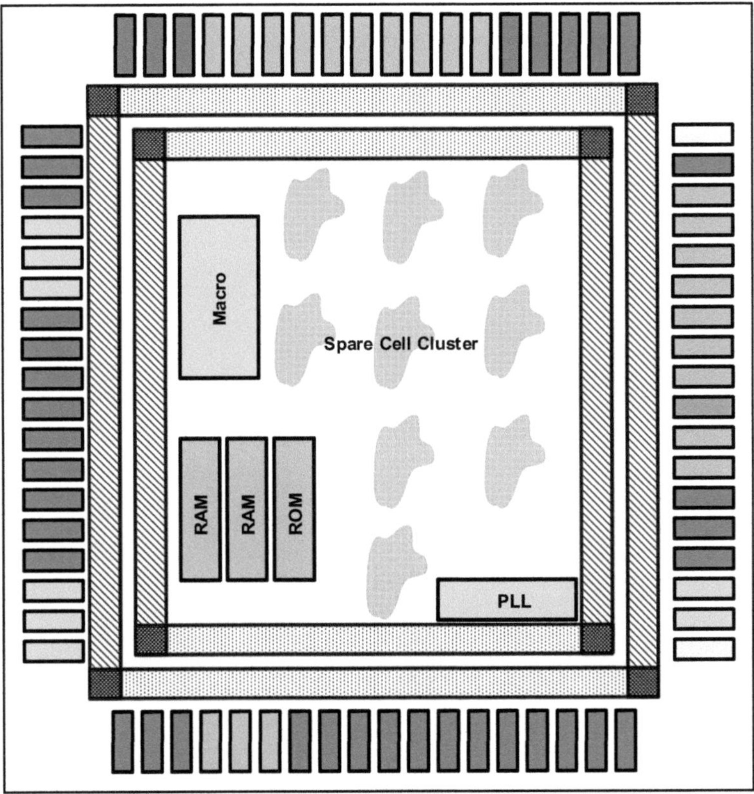

Fig. 4.7 Conceptual spare cell cluster insertion

There are two ways of inserting spare cells. One is known as a shot-gun approach which places spare cells randomly in the design area. The second way is to create pre-defined spare cell clusters from standard cells and/or metal-programmable cell ECO libraries and place them in an orderly manner in the design area as shown in Fig. 4.7.

It should be noted that spare cell clusters can be inserted in a design that has multiple power domains by inserting tap cells [3]. Tap cell insertion is the final stage of floorplanning in advanced technology.

There are two types of standard cell design. One is to have standard cells with an internal tap. In other words, the N-well is tied to the power supply (VDD), and the substrate is tied to the ground supply (VSS) as shown in Fig. 4.8.

Another type of standard cell design is when standard cells do not have their own tap (tap-less), and so are using external tap cells for N-well and substrate connections to VDD and VSS as shown in Fig. 4.9.

Having external tap cells allow standard cell designers to reduce the area of standard cell area by sharing multiple standard cells with one tap cell.

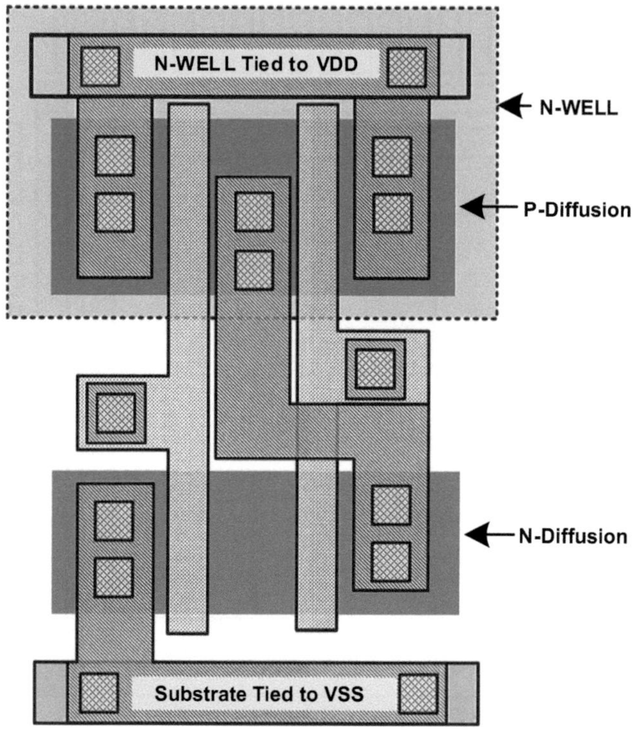

Fig. 4.8 Standard cell with internal tap

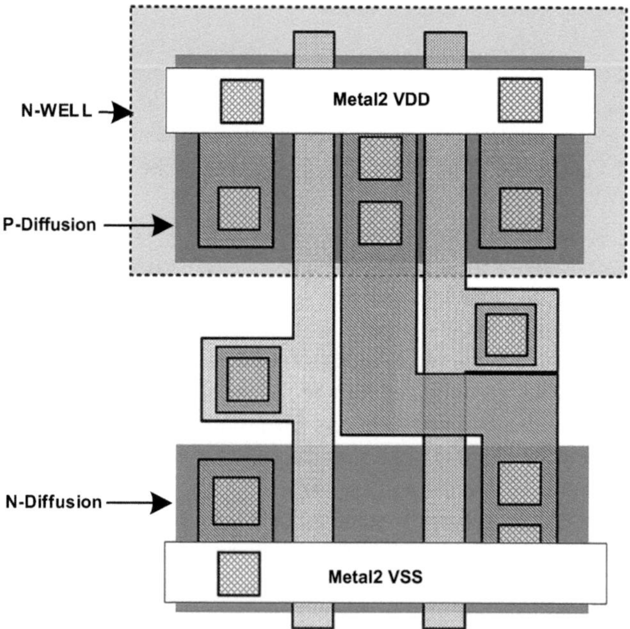

Fig. 4.9 Tap-less standard cell

In larger technology nodes, having N-well and substrate tied to VDD and VSS within standard cells was not an important area benefit due to the silicon processing design rules. For example, contact to diffusion spacing rules was of little consequence in comparison with other design rules such as routing layers.

However, for advanced nodes, the processing design rules are shrinking and by using tap-less standard cell design allows for a smaller area. To put it into perspective, in larger technology nodes, contact programmable Read Only Memory (ROM) was smaller than via programmable ones in area. For advanced node technology, it is the other way around.

It is important to note that since the internal taps are removed from standard cells, their PMOS and NMOS transistor's source and drains must be connected to VDD and VSS. This is because *metal1* is no longer available to make these connections. Therefore, the only option is to use the *metal2* layer (Fig. 4.9) in a horizontal direction like VDD and VSS of standard cells' supply rails. This changes the overall design's routing layer directions.

For non-tap-less design, the routing scheme was odd horizontal metal layers and even vertical layers. However, for tap-less design, *metal2* is the same as *metal1* for horizontal layers in order to connect power/ground supply rails to the standard cells. Having *metal2* layers in a horizontal direction has an advantage in that one can increase its width for improving the design's IR drop—provided there are no routing congestions.

The importance of using tap and endcap cells is to increase resistance between power and ground connections for N-well and substrate separately. This avoids a phenomenon known as latch-up.

Latch-up is a short circuit between power and ground which is caused by a low-resistance path between power and ground rails that are subject to an interaction between parasitic PNP and NPN bipolar transistors. This interaction is inherent in the CMOS process. The high current produced by this phenomenon damages PMOS and NMOS transistors [2].

The rules for taps and endcaps are technologically dependent. Some technologies don't require them at all because the tap connections are built in to their standard cells (generally for larger technology nodes). For advanced technology nodes that are using tap-less standard cells, a tap cell is required to be placed evenly as prescribed by the silicon foundries [3] with the endcap cells placed at the end of every standard cell row as shown in Fig. 4.10.

For designs that require tap cells, it is recommended they be placed in a matrix format with distances apart as prescribed by the silicon foundries. Figure 4.10 illustrates placement of tap (in Xum distances) and endcap cells.

During physical design verification, the DRC will flag any issues regarding tap and endcap placement. Therefore, DRC should be run early in the process.

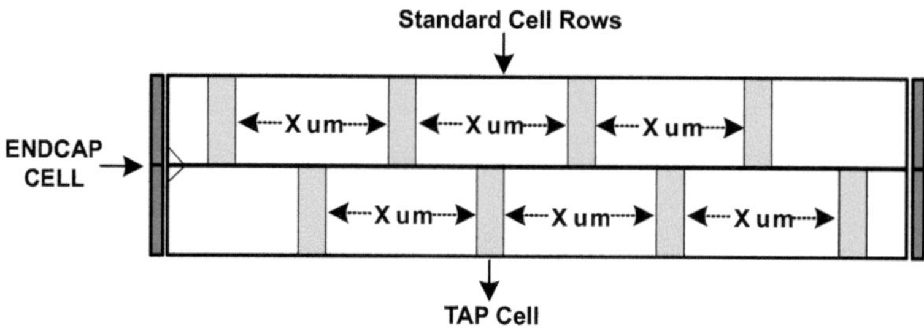

Fig. 4.10 Endcap and tap cell placement

4.2 Placement

Design settings are design-specific settings that apply to all stages of physical design and need to be modified according to the design and its associated standard cell libraries. In general, these design settings establish which standard cells should not be used (e.g., very low drive standard cells), critical nets that should be buffered and/or touched (e.g., manually routed nets), and so on. Use of multiple voltage threshold (VT) standard cells (high, standard, and low) is accomplished during the physical design optimization process. Low VT standard cells increase leakage, and, therefore, usage should be minimized.

This is especially important on advanced nodes using an ultra-low power process as the final ASIC performance could be adversely affected. While high VT standard cells have the least leakage power, its voltage threshold is very close to the supply voltage, and it has the potential of having a temperature inversion effect. On the other hand, low VT standard cells' threshold voltage is below the supply voltage; therefore, they are less affected by temperature inversion. Achieving a balance between the use of different VT standard cells is critical in ensuring the ASIC performs as designed.

The placement algorithm is programmed to fix design rule constraint requirements such as max transition, max capacitance, and high fanout nets.

Max transitions could be in the standard cell library for each cell or could be defined by the user. Max transition is used to limit long signal transition. Long signal transition reduces the cell's noise margin, increases net delay, and could cause a short circuit (PMOS and NMOS transistors to be on at same time) which increases leakage current.

Max capacitance and max fanout values are mainly user defined. They are used to reduce gate output capacitance loading for a given net (excluding clock nets) which, in turn, drives many input gates.

Although max transition, max capacitance, and max fanout can be eliminated during synthesis, it is not recommended. Rather they should be eliminated during design placement.

During design placement, max transition, max capacitance, and max fanout are fixed by inserting what is known as a buffer tree. At this stage, buffer tree insertions are not as structured as the ones used for CTS.

Although the user can set max transition value in the physical design tool, if the libraries have a value for max transition, the value in the libraries has precedence over that which was user defined.

It is important to check both best-case and worst-case libraries for their max transition values. If the values are the same, no action is required.

However, if best-case and worst-case libraries have different max transition values, the user needs to include both libraries during max transition fix.

By default, only the worst-case library is used during placement which means only those max transition violations in the worst-case library will be fixed. Because of this, the best-case library max transition violations would not be fixed and would appear during hold timing analysis in the signoff phase. Fixing these violations during that stage will require many ECOs to correct.

To prevent max transition violations from occurring, `hold_func` should be added to the setup list of the analysis view in MMMC.

In a typical physical design flow, users concentrated on the functional setup and hold timing issues. Later they would manually correct scan capture hold timing violations through the ECO process. Using MMMC flow will eliminate or minimize these types of ECO.

To take full advantage of MMMC during signoff, one needs to identify the scan clock net by finding its source and using the `do not touch` attribute. This should be done during the placement phase and before proceeding with the high fanout fix. Once high fanout fix is completed for all non-critical nets in the design (e.g., reset net), `do not touch` attributes need to be removed from the scan clock net and proceed with the high fanout fix for the scan clock using the balanced buffer insertion option (most physical design tools support this option) as shown in Fig. 4.11.

The advantage of inserting balanced buffers on the scan clock net during placement is to provide the ability of reducing scan clock skews by adjusting the inserted balance buffers at input of scan MUXs with minimal manual ECOs and without distributing functional clock tree structures in the design. This is shown in Fig. 4.12.

The placement and optimization options vary by physical design tools. However, for advanced nodes it's imperative to have options such as timing driven, clock tree insertion awareness, clustering, clock gate awareness, and leakage/dynamic power reduction.

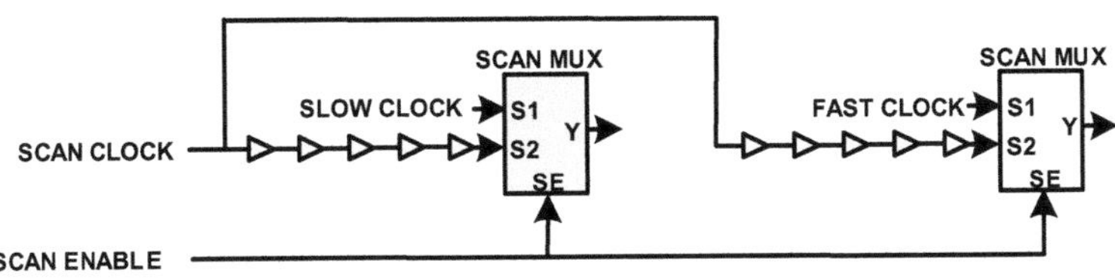

Fig. 4.11 Scan clock balanced buffer insertion

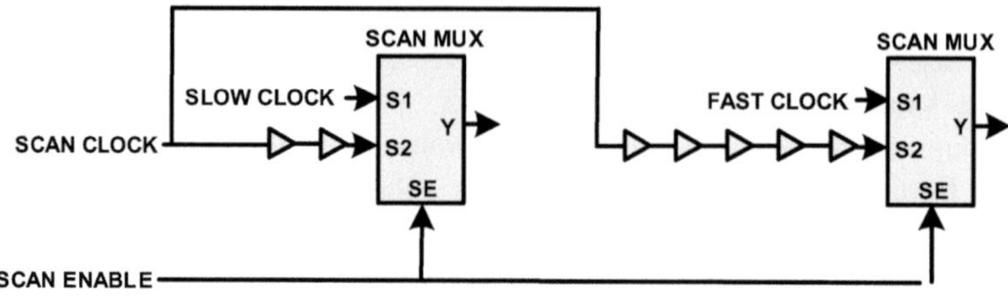

Fig. 4.12 Minimizing scan clock skew buffer adjustment

Once placement and optimization are complete, the congestion map needs to be reviewed to ensure there are no heavily congested areas.

Heavily congested areas (e.g., with utilization factor 97% or higher) not only create a routing problem (shored nets), it also has a negative impact on the design setup timing. Design setup timing must meet the design timing requirements before placement completion. In general, congested areas are created by floorplanning issues which would require refinement of the floorplan. Or congested areas may be inherent to a specific module in the design which has a very high back-to-back connectivity amount in its standard cells.

An example of high back-to-back connectivity is a synthesized small-sized ROM. While this is not a recommended practice, some RTL designers have the misconception that having a synthesized small ROM saves area (which is true at the netlist level). However, this causes routing and timing issues during placement and final routing. One remedy to resolve this type of congestion problem would be to create a region for the module with high back-to-back connectivity with a lower utilization factor in the floorplan.

After placement, there should not be any large setup timing violations (e.g., over 100 ps). If there are, they need to be investigated as to their root cause and then corrected.

The sources of setup timing violations could be floorplan and/or design constraints, such as missing case analysis, multi-cycle paths, false paths, over-constrained input/output delays, and/or clock gating cells that are not real clock gating cells (in that case, they would need to be disabled).

Design placement is a two-step process—placement of standard cells (i.e., placement mode) and optimization of those placements (i.e., placement optimization mode) according to timing constraints requirements. Placement mode options may be as follows:

- Wire length reduction
- Uniform density
- Maximum density
- Congestion reduction
- Timing driven
- Clustering
- Clock gate awareness

- Clock tree insertion awareness
- Maximum clock tree clustering fanout

Placement optimization mode options may be as follows:

- Optimization effort
- Leakage power reduction
- Dynamic power reduction
- Clock gate awareness
- Adding instances
- Use of useful skew
- High fanout fix
- Maximum wire length
- Reclaiming area

The placement and optimization options vary by physical design tools. However, for advanced nodes it's imperative to have options such as timing driven, clock tree insertion awareness, clustering, clock gate awareness, and leakage/dynamic power reduction.

Once placement and optimization are complete, the congestion map needs to be reviewed to ensure there are no heavily congested areas.

Heavily congested areas (e.g., with utilization factor 97% or higher) not only creates a routing problem (shored nets), it also has a negative impact on the design setup timing. Design setup timing must meet the design timing requirements before placement completion.

4.3 Clock Tree Synthesis

The concept of CTS is the automatic insertion of buffers/inverters along the clock paths of the ASIC design to balance the clock delay to all clock inputs [4].

To balance clock skew and minimize insertion delay, CTS is performed. Naturally, before CTS, all clock pins are driven by a single clock source and considered as an ideal net.

A typical ASIC design could contain many clock sources with different frequencies. This has made use of CTS challenging. Without efficient clock gating and clock tree implementation, design timing and power reduction cannot be achieved.

Before CTS, it's important to understand the design's clock structures and balancing requirements to have proper exceptions and to be able to construct optimal clock trees. Some prerequisites for CTS are:

- Creating non-default rules (e.g., setting large metal width and spacing for clock routing)
- Setting clock's max transition, max capacitance, and max fanout

- Selecting which cells (clock buffer, clock inverter) to use during CTS. (Although clock buffers have equal rise and fall time, to avoid pulse width violations, it is recommended to use clock inverters that balance both rise and fall time simultaneously.)
- Setting CTS exceptions

As mentioned before, CTS plays an important role in building well-balanced clock trees, fixing timing violations, and reducing the extra unnecessary pessimism in the design. The goal during building a clock tree is to reduce the skew, maintain symmetrical clock tree structure, and cover all the registers in the design and reduce power consumption.

To have well-balanced clock trees, one must understand the design clocks' latency and skew requirements provided by design constraints. However, using design constraints during CTS may cause unnecessary clock cell insertion and create timing violations. Typically, these issues arrive from clock dividers (i.e., generated clocks), unconstrained low and fast clock multiplexers (i.e., missing case analysis), default clock gating cells (i.e., non-clock gate cells), or cross domain clocks (e.g., slow clock logically OR/AND with fast clock).

To avoid unnecessary clock cell insertions and meet design timing and skew requirements, three stages of CTS are recommended. These stages are:

- Building physical clock tree structure CTS modified design constraints
- Optimizing clock tree with CTS with original design constraints
- Final clock tree timing optimization with actual design constraints

Building physical clock tree structures is the first stage (CTS1) of CTS. The objective at this stage is to build a physically well-balanced clock tree, avoiding excessive clock cell (clock buffer and/or clock inverter) insertion and to make sure the clock skew is as minimal as possible. In addition, all design timing requirements must be met. For that reason, one could use an artificial design constraints modification which is created by copying the actual design's functional design constraints. These modifications mainly are changing the clock sources by adding extra false paths.

To optimize clock tree structure, use the actual design's functional design constraints, and add additional MMMC modes such as setup and hold for CTS. In addition to selecting cells (i.e., clock buffers and/or clock inverters) to be used during CTS, one needs to provide CTS options such as max cluster fanout (i.e., the number of leaf cells in the design driven by one clock tree cell), max capacitance, and max clock transition. In addition, for timing analysis, OCV or AOCV for advanced processing nodes and Common Pessimism Path Removal (CPPR) need to be set.

During CTS stage, if there are any clock cell instance cluster (e.g., back-to-back buffers and inverters), one needs to use their physical design tool debug and diagnosis viewer (e.g., clock tree browser) to understand these clustering and resolve before continuing (uniform clock tree insertion after Clock Gating (CG) cell is desired) (Fig. 4.13).

Figure 4.14 shows a conceptual clocking browser indicating three back-to-back clock cell instance clusters. The first is for a divider Flip-Flop (FF); second is for

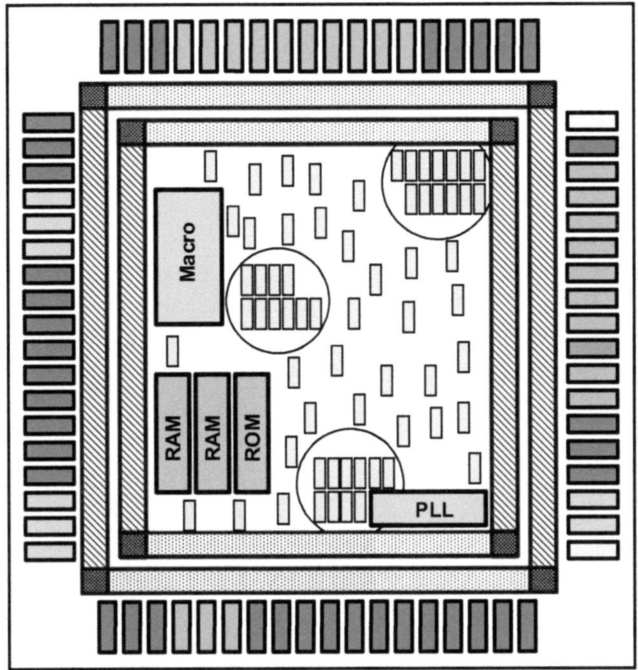

Fig. 4.13 Illustration of back-to-back clock tree insertions

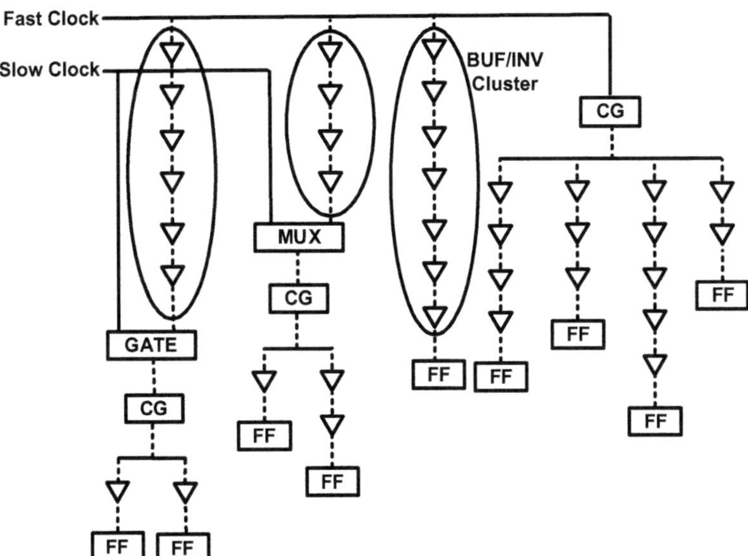

Fig. 4.14 Conceptual excessive clock cells (back-to-back) insertion

unconstrainted mux (MUX) between slow and fast clocks; and third is cross-domain clock GATE (e.g., AND/OR, etc.) between slow and fast clocks.

Once excessive clock cell insertion and large setup violations are observed, one should modify the CTS constraints by adding exceptions to certain Flip-Flop's clock port (i.e., sink pin), converting clock definition from generated to created, disabling clock gating checks on non-clock gating cells, and/or adding more case analysis and false paths to CTS constraints in order to minimize clock cell insertions and improve clock tree structures.

Based on the applied modified CTS constraints and exceptions, all leaf cells (i.e., Flip-Flop) are aligned. In other words, there are minimal overall skews and no routing congestion due to back-to-back clock cell clustering.

Removal of excessive clock cells is an iterative process so that optimal clock tree structures are achieved. This is shown in Fig. 4.15.

Removal of excessive clock cells is an iterative process so that optimal clock tree structures are achieved. This is shown in Fig. 4.15.

There may be some critical clock nets in the design that should not be touched by CTS. For those critical clock nets, one needs to insert buffer/inverter cells manually at the floorplan and set with a `do not touch` attribute. It is important to note that some physical design tools will move spare cells during CTS by default. One needs to set an option to avoid moving spare cells and ensure that clock tree's maximum fanout is set (e.g., 30). Final CTS timing optimization has two options—setup timing and hold timing fixes based on actual design constraints. Update MMMC modes for both setup and hold timing based on functional design constraints and invoke MMMC.

During CTS timing optimization using MMMC, setup timing violations are fixed first, and the hold timing violations are fixed if their timing violation fixes do not break the setup timing. This process will occur concurrently. Since setup timing violation fixes have priority over hold timing violation fixes (according to the

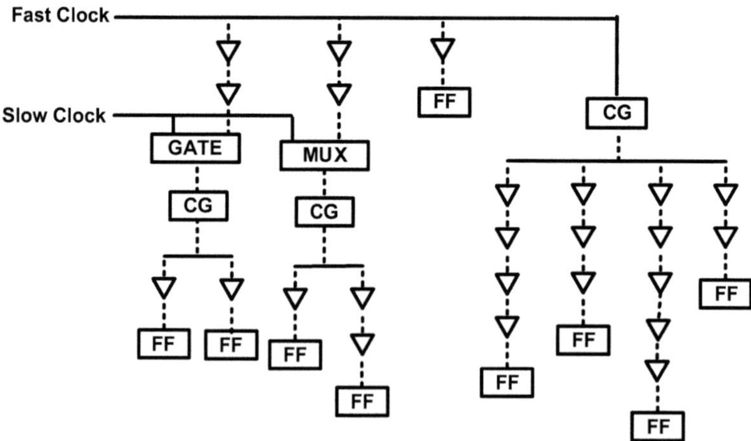

Fig. 4.15 Conceptual optimal clock tree structures

analysis view setting), there may be some hold violations that need to be fixed explicitly. But before fixing, design timing needs to be updated for both setup and hold, and timing generated for both. To fix the remaining hold timing violations explicitly, first remove `do not use` attributes for small drive buffers and delay cells, as these were set during placement and CTS.

Hold optimization modes need to be set, which include not modifying the clock structure by honoring clock domains. In addition, do not allow setup timing to be violated, and do not fix any register-to-output and input-to-register hold timing violations. A list of buffer and delay cells which will be used during hold timing violation fixes needs to be provided to ensure no excessive buffer and inverter are used.

4.4 Routing

Like floorplan, placement, and CTS flows, the routing flow will take several steps to complete. At the complication of final route, the objective is to have a physically clean route which meets all design rules, as well as timing closure, for all functional and test modes in the design. The steps for routing are:

- Initial design routing
- Design routing optimization
- Leakage and optional (SI and DFM) optimizations
- Invoke MMMC and report timing for all modes
- Apply manual ECO to remove any large timing violations
- Apply MMMC to fix remaining timing violations
- Final routing.
- ECO

The objective of initial design routing is to route and time the design, fixing setup timing violations. One should note that today's routers will not have any open nets, and therefore, open nets will not be discussed.

After completion of the initial design routing, there should not be any shorted nets. If there are any shorted nets, most EDA tools have a "search and repair" option to resolve. This option only works well in an isolated area with shorted net violations. In addition, this option may move the violations from one area to another. This process of "search and repair" may become problematic when there are multiple shorted net violations.

If there remain shorted net violations after using the "search and repair" option, one option is to delete shorted nets. It should be noted that it may require several attempts to delete all shorted net violations. Once the shorted nets are deleted, ECO and design routing needs to be performed. It is not uncommon to have many shorted nets remaining even after a few attempts of removing the violations. This could be a result of routing congestion. Often the congested area in the design results from clock net routing using Non-Default-Rules (NDR) during CTS. For that reason, remove clock routing NDRs for a specified clock net, thus relaxing the routing area.

However, caution should be exercised in using this command as it may introduce problems such as capacitance coupling in the clock routing.

Another problem in routing congestion may be caused by a lack of enough area for routing (i.e., high routing utilization). This is especially true for advanced nodes where the standard cell area utilization is high due to their small standard cell's geometry; thus there is not enough area for routing (having 12 routing layers is not uncommon for advanced nodes). To decide whether the chip area needs to be increased or routing layers added, one needs to do a manufacturing cost analysis to see which option is least expensive.

Scenic net routing, which is a byproduct of high routing utilization, impacts timing. This problem can be fixed by either increasing the chip area or adding more routing layers.

Another routing consideration is Signal Integrity (SI) which is the ability of an electrical signal to reliably carry information and resist the effects of high-frequency electromagnetic interference from nearby signals called crosstalk.

Crosstalk is due to the influence of cross coupling capacitance, which is caused by switching the signal from one net (aggressor) to the neighboring net (victim).

Most of today's EDA tools have an option to route with SI awareness. When this setting is true, the option allows the routing tool to increase the spacing between the aggressor net and victim net, so that the cross-coupling capacitance decreases as spacing increases. This then reduces the effect of crosstalk. In addition, the physical design tool inserts buffers to boost the strength of the victim net and, thereby, reduces the effect of crosstalk.

Another technique to avoid crosstalk is to place a shielding which is the ground net (e.g., VSS) between the aggressor and victim net so that the voltage will discharge through the ground net. However, this also increases the sidewall capacitance, which may in turn impact the net delay timing.

Another routing issue which may arise is glitch, or noise bump, which occurs when a transition on an adjacent signal (aggressor net) causes a noise bump/glitch on the constant signal (victim net). A noise model is required to fix glitches or noise bumps. Most of the time there are separate dedicated EDA tools for noise analysis.

After initial routing, there should not be any shorted net violations or large setup timing violations. One must time the design and generate and review the timing violations report. Once the initial routing is confirmed (i.e., no shorted nets and no large setup timing violations), the next step is routing optimization. At this stage, the objective is to address both setup and hold violations, if any. To do this, the analysis mode needs to be set for both setup and hold with OCV and CPPR options. It should be noted that the analysis mode is also set for setup and hold (`both`) timing. In addition, one needs to make sure all clocks are set in propagation mode. During routing optimization, `Delay Calculator Mode` is set for `SI Aware`, and `Extract RC Mode` is activated with capacitance coupling effects. To complete, one needs to optimize the routing. Since the design is already routed, the `Post Route` option is used during design routing optimization.

Another routing optimization step is to fix any remaining hold timing violations. If the standard cell's design library supports low leakage cells (e.g., with **LL** prefix) and they were set with `do not use` attributes because of their slow intrinsic delay, it is recommended using them for the hold timing violations fix. Delay cells in the standard cell library should also be used to fix hold timing violations. Moreover, it is recommended to use a selection of cells from the design's standard cell library for fixing hold timing violations.

Fixing remaining hold timing violations is performed with `Post Route` with `Hold` option.

Often physical design starts with using High Voltage Threshold (HVT) cells for the purposes of addressing overall ASIC design leakages. In addition, other cell types, such as Standard Voltage Threshold (SVT) and Low Voltage Threshold (LVT), are attributed with `do not use` attributes because of their high leakage. However, from a performance timing point of view, HVT cells are slower than SVT and LVT cells. Therefore, it may be necessary to use SVT and LVT cells in addition to HVT cells to fix the remaining setup timing violations. To allow EDA tools to use these cells, the `do not use` attribute needs to be removed during the fixing of setup timing violations. After removing `do not use` attributes, post route optimization needs to be performed.

The last stage of routing optimization is to minimize design leakage power. The EDA tools will replace cells with high leakage (e.g., LVT and SVT) with low leakage cells (HVT) if swapping these cells does not break design setup and hold timing. One can report timing before and after leakage optimization.

There are additional optimizations which can be performed during routing optimizations such as SI and DFM. DFM consists of a set of different physical design rules, known as "recommended design rules," regarding the shapes and polygons, (e.g., standard cell layout). DFM rules apply to spacing/width of interconnect layers, via/contact overlaps and contact/via redundancy.

DFM rules are applied to minimize the impact of physical process variations on performance and other types of parametric yield loss. For example, all different types of worst-case simulations are essentially based on a set of worst-case device parameters that are intended to represent the variability of transistor performance over the full range of variation in the fabrication process.

Changing the spacing and width of the interconnect wires requires a detailed understanding of yield loss mechanisms, as these changes trade off against one another. For example, introducing via redundancy will reduce the chance of via problems during the manufacturing process and reduces via resistance. Whether this is a good idea or not is dependent upon the details of the yield loss models and the characteristics of the ASIC design.

For advanced nodes (e.g., below 20 nm) where the manufacturing process is undergoing process improvements, it is essential to add DFM rules as much as possible if it does not impact the size of the chip area.

The initial routing and optimizations discussed thus far apply only to the main functional mode. To close all remaining setup and hold timing violations on other modes in the design, MMMC methodology should be utilized. It is important to note that in today's ASIC design, there are more than one functional (e.g., UART,

USB) and test (e.g., MBIST, SCAN) mode that can be included in setup and hold timing analysis MMMC's view lists.

Before applying post route optimization, check both setup and hold timing violations for all modes in the design. If there are any large violations (e.g., greater than 300 ps), they will need to be reduced through manual ECOs. For example, it is not uncommon to have large hold timing violations for scan capture and shift.

Fixing scan shift hold timing violations are very straightforward from the ECO point of view. Delay or buffer cells need to be added to the scan input port of the violated Flip-Flops (e.g., scan input of Flip-Flop or SI) to fix scan shift violations without cffccting any functional setup and hold timing. The reason for this is that all Flip-Flops in the design form back-to-back (i.e., scan chain) bypassing all combinational logics between them. Thus, by construction, there will be no setup timing violations among these Flip-Flops in the design.

Figure 4.16 shows an example of fixing hold timing violations during scan shift. It should be noted that two buffers were added to SI input of INST2 to resolve hold timing violations due to the scan clock skew between INST1 and INST2.

However, fixing scan capture hold timing violations is not as straightforward as scan shift ECOs.

During scan capture, the combinational logics between two given Flip-Flops are not bypassed. This is so manufacturing faults in the design will be captured.

A problem arises during the scan capture hold timing fix when scan clock is used as a functional clock. Setup and hold timing are fixed in functional mode using multiple clocks. Since a single clock is used during the scan capture hold timing fix, functional setup timing may break. Then when trying to fix functional setup timing, the scan capture hold timing may break.

During scan capture there is only one single clock (i.e., scan clock) which is used to exercise all combinational logics to determine manufacturing faults. This leads to clock skew among different functional clocking which leads to scan capture hold timing violations.

Figure 4.17 shows the conceptual issue with functional and scan clock during scan capture mode.

Adding delay or buffer cells to the data inputs of violated Flip-Flops (e.g., D input) is often used to fix scan capture hold timing violations. However, this may

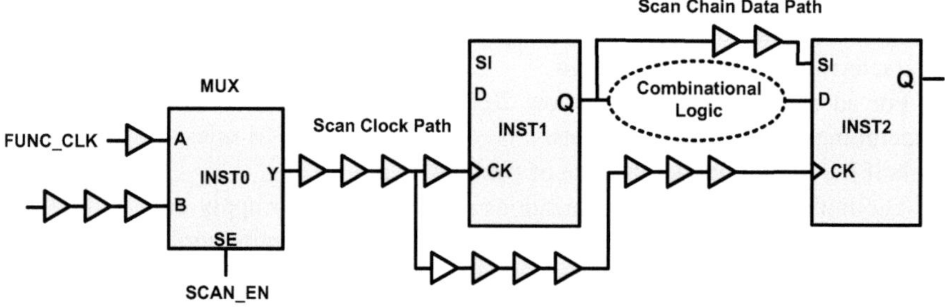

Fig. 4.16 Example of hold timing violation fix during scan shift

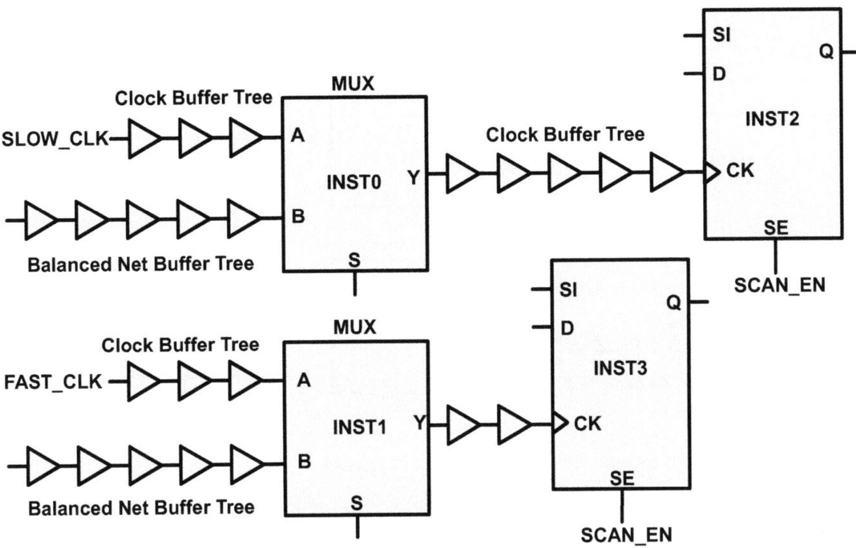

Fig. 4.17 Example of scan and functional hold timing conflicts

increase the possibility of breaking setup timing in the functional clocking modes (e.g., slow clock and fast clock).

In the typical physical design approach, this may require unnecessary manual ECOs to fix both hold (scan capture) and setup (functional) timing violations due to differences in skews impacting functional clock tree insertions.

In the example shown in Fig. 4.17, during functional mode (SLOW_CLK) there are ten clock tree buffers, and for the other clock (FAST_CLOCK), there are only seven clock tree buffers.

Although functionality is acceptable with clock skews in setup and hold timing constraints, this violates scan capture timing constraints as scan capture requires that there are no skews among all clocks in the design.

One remedy to resolve this problem with minimal impact on the functional side is to use a balanced high fanout net fix for scan clock. Using a balanced high fanout net for scan clock allows the physical designer to adjust the scan clock skews by adding buffers or delaying cells at the input of scan mux (e.g., input B of INST0 and INST1). Figure 4.18 shows how to modify buffering from a high fanout net fix with an option of a balanced net during the placement stage for scan clock muxs (i.e., their B inputs), in comparison to Fig. 4.17, without impacting functional clock tree structures (i.e., adding buffers to B inputs of INST1).

Comparing Fig. 4.17 (which is before ECO) and Fig. 4.18 (which is after ECO), one can see that three buffers were added to the input B of scan mux (INST1) to minimize scan clock skew between Flip-Flops with slow and fast functional clocks. This was done without impacting functional clock tree structures. By counting the number of scan clock paths, one can see there are no scan capture violations for

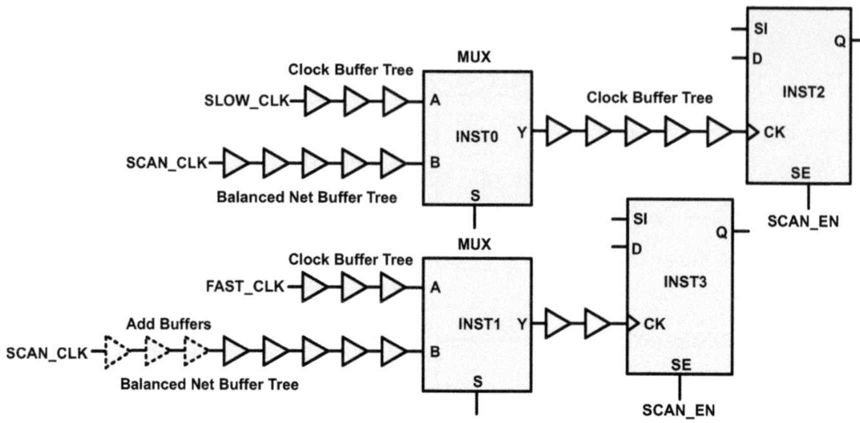

Fig. 4.18 Example of manual ECO for scan clock balancing

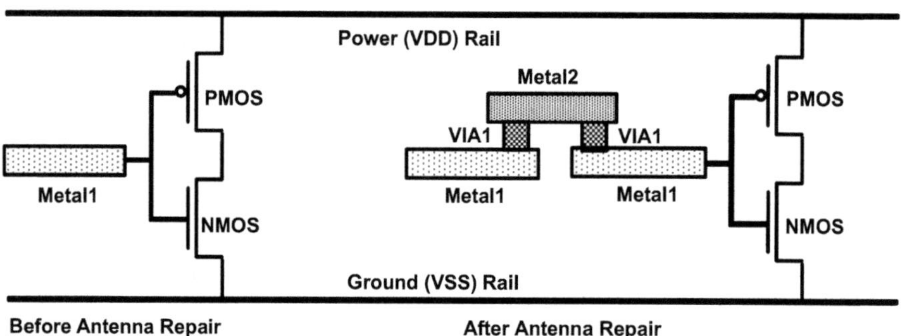

Fig. 4.19 Antenna metal repair procedure

either Flip-Flop (i.e., INST2 and INST3) as they have identical delays and minimal skews for the scan clock (SCAN_CLK).

Once all large setup and holds are removed, perform the post route optimization with post route options. The only difference between post route and timing optimization is that MMMC is activated during this step of design timing optimization.

The final steps are to export full interconnect resistance-capacitance (RC) extractions for each process corner in the design (e.g., slow and fast process), post route (often referred to as CTS netlist) for the purpose of STA, gate-level simulations, IR drop, dynamic power, and noise analysis.

After the filler cells (i.e., decap cells and metal ECO cells) are inserted, design connectivity, geometry, and antenna are verified to see if there are any issues. All issues need to be resolved.

The most common technique used to resolve antenna problems is to reduce the peripheral metal length that is attached to the transistor gates. This is accomplished by segmenting the wire of one metal type to several segments of a different metal type and connecting these various types of metals through via connections (shown in Fig. 4.19). Another way to resolve would be to insert protection diodes.

It is important to realize that these extra via insertions, done while repairing antenna violations, increase the wire resistance as the via is highly resistive. Therefore, it is strongly recommended to extract the routing parasitic after resolving all antenna violations to account for the extra resistance.

In today's ASIC designs with multimillion gates, along with the increase in design complexity, the chance of functional ECOs is equally increased. Metal ECOs play a vital role in absorbing last-minute design changes before taping out and, hence, help save millions in terms of mask cost. They also help avoid all layer mask changes in re-spins to fix bugs found during design validation. Though design fixes belong to the logical domain, physical aspects of implementation are poorly understood. Metal ECO implementation methodologies with an emphasis on mask programmable cells are discussed below.

Originally the idea of mask programmable cells was gate-array technology. The gate-array technology is an approach to the manufacture of ASICs using a prefabricated chip with components that are later connected into logic devices (e.g., NAND gates, etc.) by adding metal interconnect layers at IC foundries. Spare cells are inserted in the physical design in the form of spare modules. A spare module mostly includes a wide variety of combinational and sequential cells (e.g., inverter, buffer, multiplexer, Flip-Flop, etc.) from the standard cell library. The input of these cells is either tied to power or ground to ensure no floating outputs. This approach has some inherent probabilistic problems.

To resolve, the metal programmable ECO cells also need to be inserted during the final stage of routing. The additional programmable ECO cells are beneficial during the design ECO process. High drive strength programmable ECO cells are needed to drive long nets and avoid timing violations that are due to large capacitance loading. There are two types of metal programmable ECO cells. One is ECO filler (as shown in Fig. 4.20), and the other is functional ECO cells.

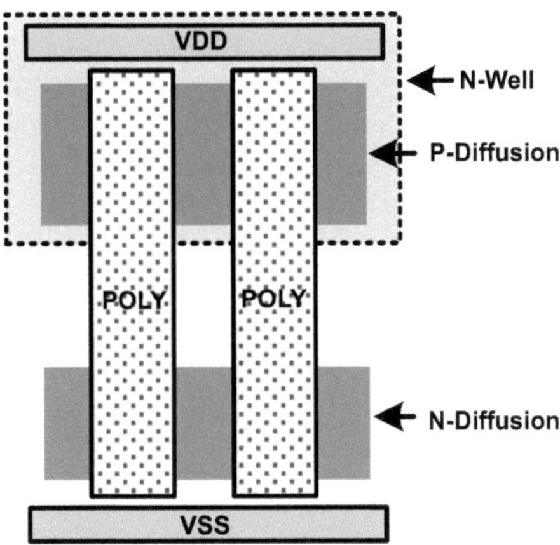

Fig. 4.20 Example of metal ECO filler spare cell

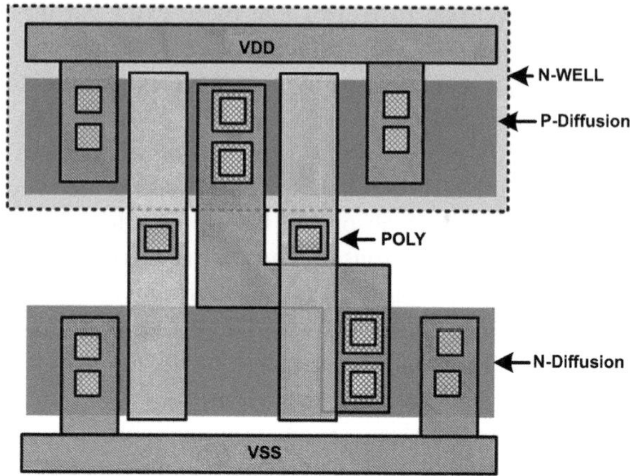

Fig. 4.21 Example of metal ECO functional spare cell

The ECO filler cells are constructed based upon the base layers known as Front-End-Of-Line (FEOL). FEOL are implant, diffusion, and poly layers. This allows any functional ECO to be performed using Back-End-Of Line layers.

Functional programmable ECO cells include a wide variety of combinational and sequential cells with multiple drive strengths realized by using width multiples for filler cells. Their cell layout has the same FEOL footprint as that of ECO filler cells. The only difference is that the functional ECO cell will use ECO filler FEOL layout and has contact connections to poly, diffusion, and *metal1* layers for internal connections to construct a functional gate. Figure 4.21 shows a functional ECO cell with the same footprint as the ECO filler cell shown in Fig. 4.20.

4.5 Summary

This chapter explains various aspects of the physical design process, data preparation, and ASIC physical design and floorplanning alternatives.

In the area of data preparation, several timing and design constraints, and their impact on the quality of the ASIC physical design floorplan, placement, CTS, and routing are discussed.

Also, the fundamentals of different floorplanning styles, mainly flat and hierarchical, and their advantages are addressed. Likewise, a review of basic floorplanning techniques is offered.

Construction of power domains, either multi-voltage or power gating, is discussed. Level shifters (low-to-high and high-to-low) are used in the construction of a multi-voltage domain.

For the power gating style, header cells are used to construct the header cell ring-chain. The header cell ring-chain includes the header, header filler, outer corner, inner corner, and power protection (CLAMP) cells. In conjunction, a sample of CPF rule descriptions is included.

It should be noted that floorplanning, placement, CTS, and routing style depend upon many factors such as the type of ASIC, area, and performance and rely heavily on one's physical design experience.

References

1. Naveed Sherwani, *Algorithms for VLSI Physical Design Automation, SECOND EDITION*, Kluwer Academic Publishers, 1997
2. Khosrow Golshan, *Physical Design Essentials, an ASIC Design Implementation Perspective*, Springer Business Media, 2007
3. Kurt Keutzer, *Closing the Power Gap between ASIC & Custom: Tools and Techniques for Low Power Design*, Kindle Edition 2007
4. Khosrow Golshan, *The Art of Timing Closure, Advanced ASIC Design Implementation*, Springer Nature, Switzerland AG, 2020

Chapter 5
Design Verification

*For a house to be successful, the objects in it must communicate
with one another, respond and balance one another.*
–Frank LIoyd Wright

Verification is the final phase of any ASIC physical design before submission of the device for fabrication. This phase focuses on the testing of functional correctness and design manufacturability.

As ASIC designs are becoming more complex with respect to increased circuit density and advanced semiconductor process, verification of these devices is becoming more difficult. The main objective of verification is to ensure the functionality of ASIC designs and to minimize the risk of creating a design that cannot be manufactured.

Comprehensive verification is an iterative process and has a time complexity that is linear with respect to the size of the design and sublinear with respect to memory usage. One of today's ASIC design implementation challenges is to manage the iterative verification process, in terms of the time required, and to be able to improve throughput by means of automating the process of verification.

ASIC design verification is becoming an increasingly important phase of any physical design as design complexity increases and traditional verification tools have proven to be insufficient. A comprehensive ASIC design verification process consists of three phases. These verification phases are:

- Functional
- Timing
- Physical

K. Golshan, *ASIC Design Implementation Process*,
https://doi.org/10.1007/978-3-031-58653-8_5

5.1 Functional Verification

Functional verification is performed during the early ASIC design implementation stage. Functional verification uses mixed tools and technologies for performing logic simulation, simulation acceleration, assertions, in-circuit emulation, and hardware/software co-verification.

Early-stage functional verification consists of three sequential tasks—modification, test, and evaluation. Standard functions and algorithms are used to verify the design behavior and compliance with the specification.

Once the design requirement is satisfied, the behavioral design description (i.e., RTL) is transformed into the structural domain, or gate-level description, by virtue of logic synthesis. Upon completion of the logic and layout synthesis, functional verification is performed against the behavioral RTL pre-layout and post-layout structural description (netlist) for design validation. The final design verification is accomplished by utilizing either simulation-based verification, rule-based verification, or both.

During simulation-based verification, the same stimulus is applied to the design behavioral description as to the final gate-level description, and their responses are compared and evaluated as shown in Fig. 5.1.

Although functional verification through simulation constitutes the current state of the art, this method has two main disadvantages. One is that simulation-based verification requires a long execution time and, therefore, often becomes impractical when ASIC designs are very complex and large. The other disadvantage is there may not be a comprehensive set of stimuli available to validate the entire design, forcing the designer to rely on some method of random stimulus generation for full design coverage.

To address the growing functional verification requirements and problems associated with simulation-based verification, various techniques have been suggested. Among these techniques, rule-based functional verifications have been adopted for very large and complex ASIC designs.

The most widely accepted rule-based functional verification techniques are assertion-based and formal-based functional verification. Assertion-based functional verification uses an assertion library containing assertion macros that are used to verify specific design properties. Assertion macros are expressions that, if false,

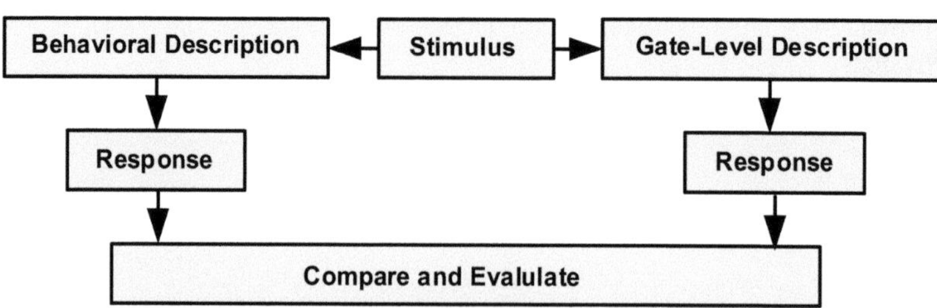

Fig. 5.1 Simulation-based functional verification flow

indicate an error and are instantiated within RTL code that provides diagnostic information during the functional verification.

The levels of visibility and controllability provided by assertion-based verification are becoming ever more important as the size and complexity of ASIC design increase.

Although assertion-based verification has been used for many years (specifically in HDL for functional simulation), it has more recently been used as an enabler of much more efficient functional verification methods.

One of the important aspects of assertion-based verification is to be able to reuse the assertion library for multiple verification tools with consistent specification mechanisms. Recently, there has been an Open Verification Library (OVL) initiative that focuses on defining the assertion library for open-source standardization. Thus, the idea of assertion-based verification goes beyond functional capability and encompasses usability and methodology.

In contrast with assertion-based functional verification, formal-based functional verification uses equivalency and model checking to verify the RTL versus the structural (gate-level netlist) description. There are two methods of formal-based functional verification that are frequently used, model checking and equivalence checking.

In model checking, in order to formally verify a design, the design must first be converted into a verifiable format that obtains a complete Finite State Machine (FSM) specified as a set of interacting components. Each interacting component has a finite number of properties or states, and the states and transitions between states constitute the entire FSM. Most of the time, design FSM conversion is accomplished by flattening the hierarchical design description into a single network and computing the outputs from the inputs by the network that consists of logic gates, sequential elements, and their interconnections.

The next step in model checking verification is to convert the flattened design description into a functional description that represents the output and the next-state variables as functions of the inputs and current state (creating design FSMs). Once FSMs are created, the formal verification algorithm traverses through each FSM in a recursive fashion and evaluates the values that are stored in all sequential elements for their true values. The network satisfies the algorithm if stated values are true for all states of the network.

Equivalence checking is another method of formal-based functional verification. Formal verification, or Logic Equivalence Check (LEC), refers to a technique that mathematically verifies that two design descriptions being verified are functionally equivalent. LEC provides a formal proof that the output from synthesis and the physical design tools match the original RTL code. All of this is done without having to run a single simulation.

An ASIC design passes through various steps, such as synthesis, physical design, design signoff, ECO, and numerous optimizations, before it reaches production. At every stage, one needs to make sure that the logical functionality is intact and does not break because of any of the automated or manual changes. If the functionality changes at any point during the process, the entire chip becomes faulty. In general, LEC is a three-phase process—setup, mapping, and compare.

At the setup phase, the LEC tool reads in two design descriptions—golden and revised. The golden design description could be the design's RTL or the design's synthesized (e.g., pre-layout) netlist. The revised design description is based on the design's routed (e.g., post-layout) netlist.

In addition, the execution of LEC requires a list of libraries (e.g., standard cells and macros) and formal verification constraints. These verification constraints can include constraints such as ignoring specified cells, some scan connections, and input/output pins.

In transition from the setup phase to the mapping phase, both golden and revised design descriptions are flattened and mapped to the key points. The usual key points are:

- Registers (Flip-Flops and latches)
- Primary inputs and outputs
- Floating signals
- Assignment statements in the revised design description
- Black boxes

During the mapping phase, the LEC tool maps the key points (both golden and revised) and compares them. If there are differences, a report is generated.

In general, there are two types of mapping. One is name-based and the other is non-based mapping. Name-based mapping is used for gate-to-gate mapping such as the pre-layout and post-layout netlists. The non-based mapping is useful when golden and revised design descriptions have completely different names (e.g., instance and nets).

The key points that are unmapped during this stage are:

- Extra (the key points that are present in the golden or revised design)
- Unreachable (the key points that are not observable such as primary inputs)
- Non-mapped (the key points that are observable without corresponding instances/nets)

In the compare phase, the LEC tool compares both golden and revised design descriptions' key points to determine if they are:

- Equivalent
- Non-equivalent
- Inverted equivalent
- Aborted

The LEC verification flow is as follows (Fig. 5.2):

It is important to note that blocks with critical logical failure found at the time of signoff will cause production delays of the ASIC. At times, the logical connectivity is broken while doing manual fixes or timing ECOs. Since use of MMMC methodology minimizes manual ECOs, one can see that by using MMMC, the formal verification phase is favorably impacted.

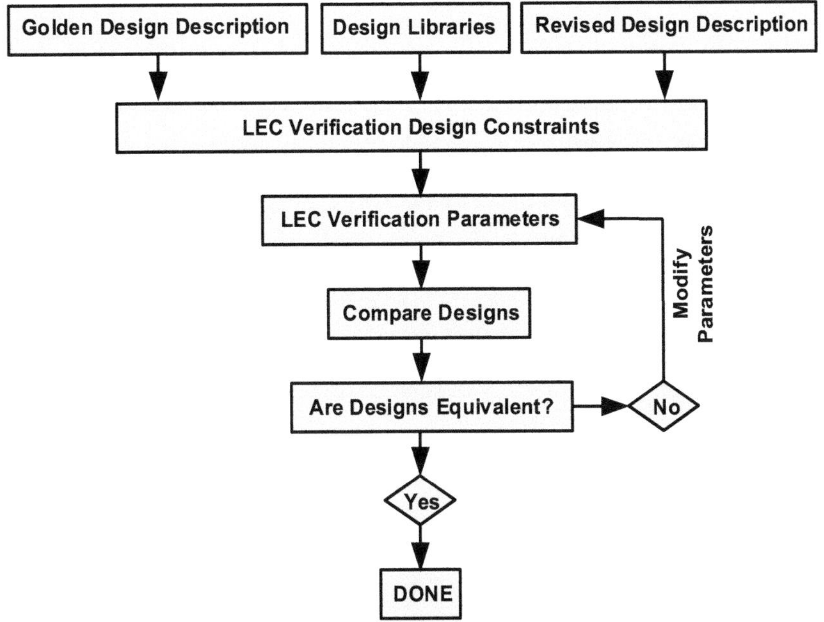

Fig. 5.2 Formal verification (LEC) flow

5.2 Timing Verification

The objective of timing verification is to ensure that the ASIC design meets all timing requirements. Two methods of timing verification are in use [1]:

- Dynamic simulation
- STA

Dynamic simulation is performed either at the transistor level or (logic) gate level and supports a wide range of design styles while revealing the impact of gate delays and parasitic effect on functionality and performance.

STA is another important part of ASIC design verification. However, there is an age-old problem that continues to challenge engineers in that STA has some complications with optimization. There is a miscorrelation between implementation timing and signoff timing when the physical design and STA tools use different timing engines.

The only alternative to address this problem is to margin the design during the physical design phase so as not to incur violations during signoff timing which uses golden STA tools. Before performing timing analysis, the extraction files (SPEF) and netlist (Verilog) need to be exported from the physical design tool.

To understand the nature of these timing mismatches between physical design and golden STA tools, one needs to report timing for the same starting and endpoint for a given path that has violations (i.e., setup/hold with negative timing slack) and make sure the violations for both tools are identical.

If there are different paths with the same start and endpoint, the conclusion is not valid. If the result is not valid, change uncertainty values (i.e., setup and hold) to compensate for the difference, provided both paths use the same design constraints.

Once physical design and golden STA tools are correlated, if there is any hold and/or setup timing violations, ECOs are applied to fix.

With power being one of the most critical design metrics for advanced node technologies, adding excessive gates to ensure minimal violations at signoff is not a proper trade-off. This leads to design power increases and may require the ASIC area be increased. This, in turn, would require the entire physical design flow to be repeated.

The iterations required to close timing differences between physical design and signoff timing are further complicated by the fact that signoff STA timing engines are not physically aware (i.e., typical signoff ECO flow). Placement of inserted buffers, inverters, and upsizing/downsizing cells is left as a post-processing step for ECO generation.

Highly utilized designs inherently have a lack of vacant space. Therefore, the placement of new or ECO cells (e.g., buffers and inverters) can be dramatically different from what is assumed by the optimization algorithms. This results in a significant mismatch between the assumed interconnect parasitic during optimization and the actual placement and routing of the ECO cells in the design.

There are potential uncertainties involved in ECO cell placement which affect assumptions made in a typical signoff ECO flow. For instance, ECO cell placement may impact the timing of paths that have already met timing.

Not only is the impact on timing unknown, it is also impossible to tell which paths will be affected by the movement of ECO cells. In a highly utilized design, the ECO cells may not be placed where they should be placed. This may result in what previously may have been timing clean and could potentially have many violations after routing and placement legalization of the inserted ECO cells. These are just some of the issues in using the typical design flow. To address, the MMMC flow is recommended [2].

It should be noted that timing optimization during the ASIC design timing signoff phase is limited to a logical view of the ASIC design and, therefore, makes assumptions about the placement of inserted cells that vary radically from the actual physical design. Understanding the location of cells, the vacancies available, and the topology of routes during the timing optimization process results in design changes that predictably match the reality of design implementation. This capability not only produces better quality ECOs but also minimizes the impact to paths that already meet timing. Therefore (in addition to use of the MMMC flow), it is strongly suggested that both physical design and STA engineer be one engineer. One engineer is responsible for the entire process, from implementation to signoff for advanced node design. Having one physical design engineer for the entire process provides for better results.

Today's design closure requires the physical design tool's timing engine to operate as the same timing engine for the signoff STA tool. In this way "what-if" analysis can be performed before committing changes to an ECO. Some solutions on the market start with signoff results but lack the integration to perform "what-if" analysis by iterating through the signoff tools. Having all design timing violations fixed during physical design (i.e., after final) routing using MMMC methodology reduces the number of ECO iterations. This allows the last piece of the puzzle to achieve timing closure convergence (i.e., physically aware). Using physical design tools for timing closures (i.e., physically aware) reduces uncertainty that occurs during legalized placement and routing of an implemented ECO. It's important to note that using golden STA tools (i.e., non-physically aware) for final design timing signoff should not be used for ECO implementation. It should be used only for checking the QoR.

Another consideration is computing power. One of the issues for ASIC design of advanced nodes is the increased physical design and verification run time. For example, the physical and timing analysis engineer must run every mode-corner combination. The quickest way to arrive at a consolidated answer is to distribute the mode-corner combinations across a computer farm and run them in parallel.

There may be some opportunity to bind some operating corners by focusing on extreme corners, but with temperature inversion affecting cell delays, and with phase mask shift uncertainty at 20 nm and below, one can never be sure that the corners have caught all violating paths between the expected best- and worst-case conditions. Today, exhaustive timing analysis and optimization are the recommended approach to ensure that a design works under all possible operational conditions. Once one has distributed all the mode-corners to individual servers, the bottleneck becomes the pure processing time required for a single mode-corner. Physical design, verification (i.e., timing and physical), and parasitic extraction tools must be architected to leverage multi-core processing.

Today's EDA tools scale reasonably well up to four cores and provide practical scalability up to eight. But while portions of the timing flow today can scale beyond eight, the performance speedup beyond eight has a negligible return when considering the overall timing optimization flow.

Scalability with an increasing number of cores is a must-have for analyzing large, advanced ASIC designs, especially with servers containing up to 16 cores. Because scalability is largely dictated by the percentage of processing steps that are multi-threaded, it is not enough to claim that only portions of the timing optimization and extraction flows are threaded.

For example, all steps, from importing the design to reporting or generating output files, must be multi-threaded. This is because the capacity of timing optimization, which is typically in the realm of implementation (e.g., physical design and STA) tools, is generally limited in capacity when optimizing over multiple views.

Most EDA vendors recommend optimizing a subset of mode-corners so as not to exceed the capacity limitations of their EDA tools. When optimizing timing across more than 100 mode-corners, it is critical to have the capacity to construct a common timing graph that represents a composite picture of all modes and corners within the ASIC design.

Reduction of mode-corners must not lead to a loss in accuracy or missed failing endpoints. Otherwise, the yield of timing fixes becomes inaccurate and has a negligible benefit. Miscorrelation between implementation and signoff engines increases the number of ECO iterations needed to close timing of a design. Users have gone through great lengths to build custom solutions that leverage timing attributes and reports from the signoff engine for the purpose of timing optimization.

With today's complex ASIC designs that have many modes and corners, it is imperative to have a common script for generating all STA scripts for all design implementation projects. This ensures the correctness of the STA process.

STA is another important part of ASIC design verification. However, there is an age-old problem that continues to challenge engineers in that STA has some complications with optimization. There is a miscorrelation between implementation timing and signoff timing when the physical design and STA tools use different timing engines.

Although gate-level simulation can handle large designs, it does not provide as much accuracy as transistor-level simulation. But with designs comprising several million transistors, creating vector sets (stimulus) that cover all functional behaviors in an ASIC device for transistor-level simulation is impossible. Because of that, gate-level simulation has been commonly adopted in the past for timing verification.

During the past decade, however STA has emerged over dynamic simulation as the preferred method of timing verification. STA eliminates the need for vector sets and provides for exhaustive timing analysis across all input and output paths in the design. Regardless of which method is chosen for timing analysis, both dynamic simulation and STA require that logic gates and wires are provided with path delays within the design.

The gate delays are provided by the macros and standard cell libraries. Since the timing models in the library give the gate delays, one of the key aspects in path delay computation is the wire delay calculation based on the ASIC final parasitic information such as effective capacitance and resistance for each net in the design.

Gate-level simulators rely on a higher level of abstraction in comparison to switch-level simulators by replacing the low-level devices, such as transistors, capacitors, and resistors, by logic functions (e.g., AND, OR). Effective use of logic functions or gates allows very complex designs to be easily described and subsequently simulated at the gate level rather than at the switch level.

Most of these types of simulators use events during the simulation. Events are defined as incidents that cause the system to change its state during a distinct unit of time during simulation. An occurrence of these events provides an effective dynamic environment in which a circuit is simulated. Although this class of simulator has been used for timing verification for many years, there are only a few that are golden for ASIC timing sign-off by ASIC foundries and designers. Among these golden simulators, Verilog gate-level netlist simulator is one of the most widely accepted gate-level simulators that is utilized for functional timing verification.

To use Verilog gate-level netlist simulation after completion of the physical design for the purpose of timing analysis, one must export the post-layout structural netlist along with the Standard Delay Format (SDF). The SDF file incorporates the actual timing information such as gate delays, wire delays, and timing checks into an abstracted format. The SDF annotator via Programming Language Interface (PLI) parses the SDF in the simulator.

During parsing of the SDF file, the annotator may report errors that are either fatal or nonfatal. Fatal errors cause the SDF annotator to stop, whereas nonfatal errors will cause the SDF annotator to skip the actions that cause the errors. The root cause of nonfatal errors is mainly found to be due to inconsistencies found during annotation, such as when a condition specified in the SDF file cannot be matched with the Verilog library that is used. These types of errors need to be resolved prior to simulation.

The SDF format uses a keyword (known as a construct) to store timing information. These constructs are typically related to:

- Delays such as interconnect, ports, and devices
- Timing checks for hold, setup, recovery, width, and period
- Timing constraints
- Delay type such as absolute and incremental
- Conditional and unconditional path delays and timing checks
- Data type or library
- Scaling, environmental, and technology

For post-layout timing simulation, the SDF file contains path constraints such as interconnect, related gate delays, and timing check constructs that are exported from the physical design tools. Figure 5.3 shows a circuit consisting of two instances connected in a series, and its corresponding SDF is illustrated in Fig. 5.3.

Inspecting the illustration of the SDF file as shown in Fig. 5.4, the file consists of design specific, or header entries, and delay entries that specify the delay values associated with interconnection between devices and ports.

The header contains information relevant to the entire design environment and its operating conditions such as voltage, process, and time scaling factors.

The delay entries that are denoted by the DELAY keyword specify the delay values associated with interconnections between devices and ports and the gate delays.

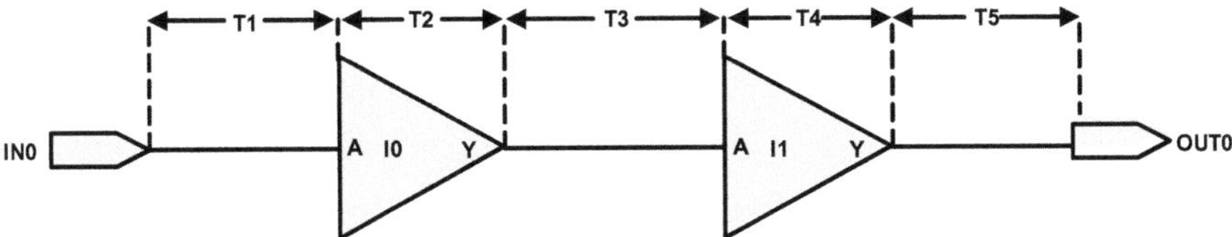

Fig. 5.3 Circuit of two connecting instances

Fig. 5.4 SDF corresponding to two connecting instances circuit

```
(DELAYFILE
  (SDFVERSION "2.1")
  (DESIGN "basic netlist")
  (DATE "Thu Dec 22 12:12:24 2004")
  (VENDOR   "None")
  (PROGRAM  "None")
  (VERSION  "1.1.0")
  (DIVIDER / )
  (VOLTAGE 1.2:1.2:1.2)
  (PROCESS 1:1:1)
  (TEMPERATURE 25:25:25)
  (TIMESCALE 1ns)

(CELL
  (CELLTYPE "basic"
  (INSTANCE )
  (DELAY
    (ABSOLUTE
      (INTERCONNECT IN0 I0/A  (T1:T1:T1) (T1:T1:T1))
      (INTERCONNECT I0/Z I1/A  (T3:T3:T3) (T3:T3:T3))
      (INTERCONNECT I1/Z OUT0  (T5:T5:T5) (T5:T5:T5))
(CELL
  (CELLTYPE "buffer")
  (INSTANCE "I0")
  (DELAY
    (ABSOLUTE
      (IOPATH  A Z  (T2:T2:T2) (T2:T2:T2))))))
(CELL
  (CELLTYPE "buffer")
  (INSTANCE "I1")
  (DELAY
    (ABSOLUTE
      (IOPATH  A Z  (T4:T4:T4) (T4:T4:T4)))))))))))
```

Interconnect construct represents the actual delay of the wires between instances that have INTERCONNECT syntax. The timing relation between an input port and an output port of an instance is considered gate delay with IOPATH syntax.

The delays that correspond to interconnect are specified as INTERCONNECT and are followed by the source and destination of wire connections and delay values (low-to-high and high-to-low) between output and input ports. These delay values are calculated by some method of estimation such as Elmore or Asymptotic Waveform Evaluation (AWE).

The delay that represents a timing arc on a path from an input to an output port is specified by IOPATH and is followed by unique input and output ports on an instance and their corresponding low-to-high and high-to-low path delays. These path delays are determined by using input transition and output effective capacitance loading. Since effective capacitance is a function of the driver's internal resistance and computation of the k-factor, one must make sure these values are properly calculated.

For both statements, INTERCONNECT and IOPATH, the values that are enclosed by parentheses represent minimum, typical, and maximum process, temperature, and voltage. If these values are different, one must specify which value is to be annotated.

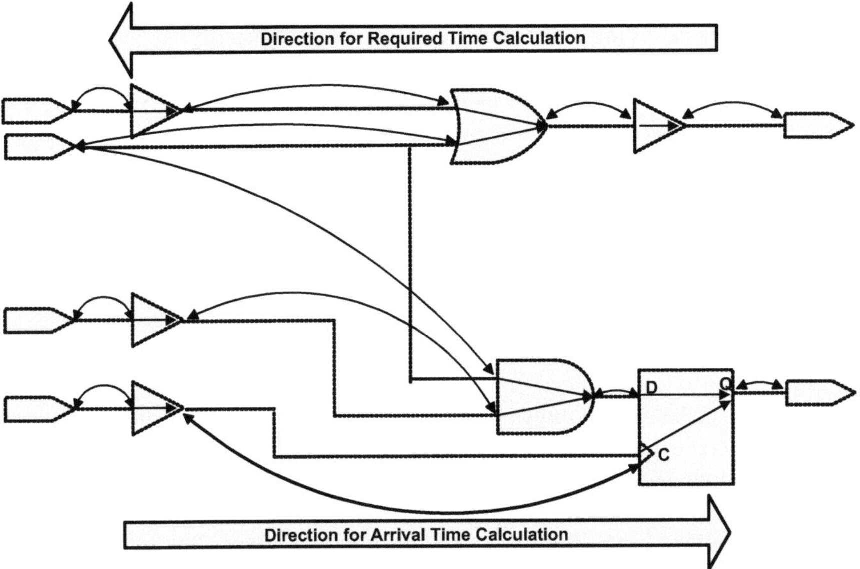

Fig. 5.5 Timing graph representation

Another option is to generate all these values for the same given minimum, typical, and maximum condition. For instance, in the example shown in Fig. 5.5, the values of **T1**, **T2**, **T3**, **T4**, and **T5** can be based on minimum, typical, and maximum conditions using three different SDF files for each circumstance or by using a single SDF file with three different delay entries.

The cell entries that begin with the `CELL` keyword identify specific design instances that are specific to a design, library, or type, as well as their associated timing checks that determine the constraints between signals and how the signal can change in relation to other signals. For instance, the timing check associated with sequential elements can be `SETUP`, `HOLD` (with format that is like `INTERCONNECT`), and `IOPATH`.

Logic simulation has been part of ASIC design validation methodologies for a long time. However, if every cloud has a silver lining, then the cloud that affect logic simulations of large and complex ASIC is lack of comprehensive stimulus that exercises all paths in the design, simulation performance (e.g., event computation), and memory requirements (e.g., event storage). Because of these limitations, STA has emerged as the preferred method for timing verification.

STA eliminates the need for comprehensive stimulus and provides exhaustive analysis of timing paths by computing the timing arcs between all input to output paths in the design. STA is considered an efficient method for performing case analysis that ensures all critical paths meet their timing requirements for a given design. Owing to the efficiency of STA, it is used as a core engine in many layout synthesis tools, allowing one to analyze timing requirements as more information becomes available during different stages of the physical design.

In principle, STA uses `arrival time` (T_a) which represents the latest time a signal can transition at a node due to the change at the circuit primary inputs and its forward propagation to the primary outputs.

In a similar fashion, STA uses `required time` (T_r) which represents the latest time a signal can transition at a node based on timing constraints that are propagated from the circuit primary outputs to the circuit primary inputs. Using both arrival and required time, the relation that is given by Eq. (5.1) must hold for the circuit to function properly.

$$T_r - T_a \geq 0 \qquad\qquad\qquad (5.1)$$

STA tools make use of timing graphs to compute the values of T_a and T_r. The timing graph is created automatically and is a directed graph where each port is represented by a node and nets connecting these ports by an edge. The start and ending points of interconnection are determined by the netlist, and internal edges of each cell are determined by their timing arcs defined in the library.

Timing graphs are very compact representations of the timing constraints and, at each node, have an associated positive or negative timing slack depending on the value of arrival time and required time.

A positive timing slack indicates that the node is earlier than the timing constraint and implies that the arrival time can be increased without affecting the overall delay of the design.

Negative slack means that the node is late and implies that the arrival time must be decreased to meet the required performance. One should note that the total number of timing slacks needed to completely constrain the timing of a given ASIC design is proportional to the number of sequential elements that are used and tend to grow exponentially as the design size increases. A basic timing graph is shown in Fig. 5.5.

Timing analysis is carried out in an input-dependent manner, and the objective is to determine all worst-case delays of an ASIC design over all the possible input combinations. This is an exhaustive calculation, and, if the circuit functionality is not considered, timing analysis tools may report some paths that are not meaningful. In order to improve efficiency and accuracy of STA, one must include the circuit states that do not exist in the actual application to distinguish whenever a long path found in the circuit is a true path or a false path.

A true or false path is defined as a set of connected gates that start from one input and end at another output. Creating all the false paths for an ASIC design can be as challenging as developing a set of comprehensive stimuli for logic simulation. Usually, an iterative approach is applied during physical design to identify all possible false paths. At each iteration, timing analysis is performed to identify a new set of false paths until all actual critical paths are optimized.

The iterative process of identifying false paths is very time-consuming as the number of false paths identified may become very large. Therefore, a technique that is effective in identifying the false path is needed to reduce the cycle of this iterative process. For this reason, many approaches for determining false paths automatically have been developed or proposed.

One of the methods used for automatic false path generation relies on a process of sensitization. In this process, a so-called false path cannot be sensitized in terms of controlling logic value. The controlling logic value is a single input value applied to an input of a gate that forces the output of the gate to a known value independent of the other inputs to the gate. For instance, applying the logic value of zero to an input of AND gate or applying the logic value of one to an input of OR gate will be considered a controlling logic value. Thus, a gate is sensitized if a signal transition can propagate through it from a particular input to the output as long as the other input has a non-controlling value such as logic value of one at input of AND gate or logic value of zero at input of OR gate. These sensitization-specific conditions are also referred to as sensitization criteria. During the sensitization process, the side inputs provide the controlling logic values associated for a given path, as illustrated in Fig. 5.6.

The sensitization criteria can be specified statically or dynamically. For a sensitization criterion to be static, there must exist a set of input vectors that create non-controlling static to all side inputs along a path as shown in Fig. 5.7.

The objective of dynamic sensitization criteria is to generate a set of input vectors such that when they are applied at different times, they will produce non-controlling logic values at side inputs when the propagating transition signals arrive at a particular gate.

Depending on the timing for a signal transition from logic zero to one, or logic one to zero, from side input one and two, a path could be considered as true or false. Due to these types of timing dependencies, dynamic path sensitization is a very complex problem, and there is much ongoing effort to simplify the process. Figure 5.8 shows the concept of dynamic sensitization.

It is key to realize that for the determination of sensitization criteria, the path delays are not under- or overestimated. In addition, one must accommodate imperfect timing information such as process parameters and operating conditions.

It is essential to have a set of comprehensive constraints and clock scheduling to properly analyze timing paths. The design constraints consist of input, output, and input to output delays with adequate timing margins to compensate for variations in design conditions such as process and temperature.

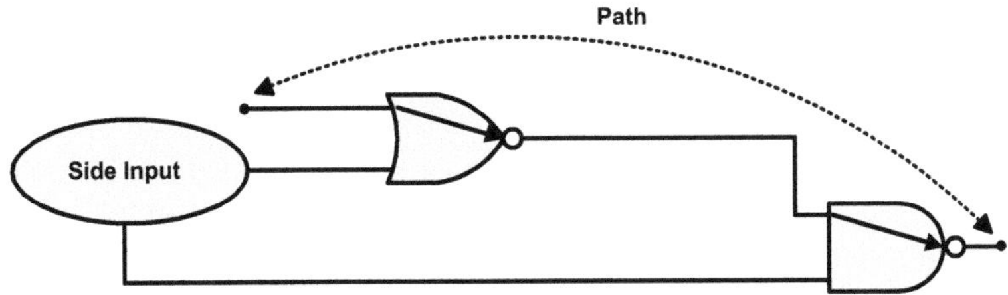

Fig. 5.6 Input side and path illustration

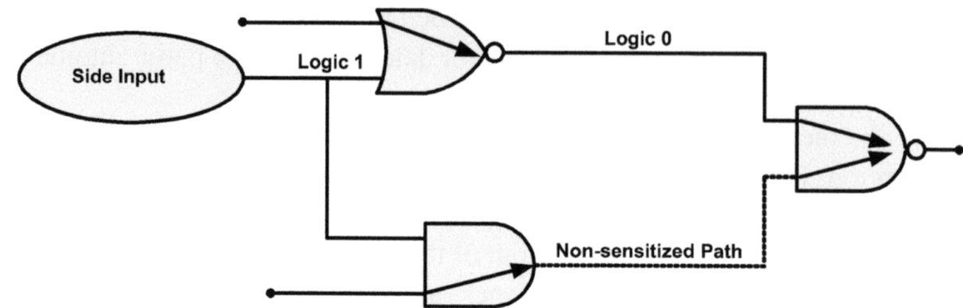

Fig. 5.7 Static sensitizations

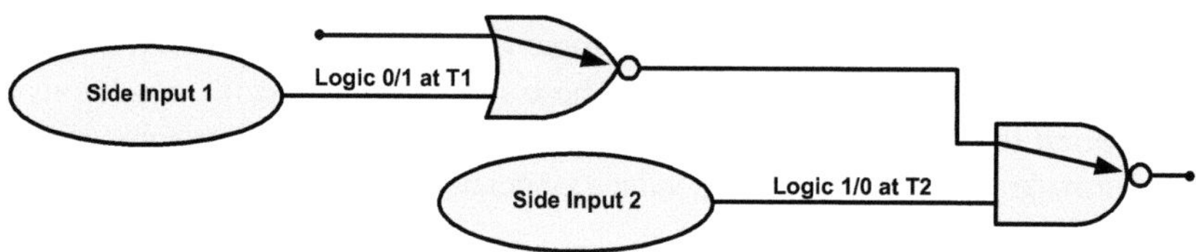

Fig. 5.8 Dynamic sensitization

During timing analysis, one must make sure that these delays do not violate any design `setup` or `hold` timing requirements with respect to the clock interval. The setup constraint specifies the time interval before the active edge of the clock, and data must arrive before the interval. The hold time constraint specifies an interval after the active clock edge, and data must be stable in the interval.

From a timing analysis point of view, several timing paths are a matter of interest. These types of timing paths are:

- Input-to-register
- Register-to-register
- Register-to-output
- Input-to-output

These various timing paths are illustrated in Fig. 5.9.

Most timing analysis of an ASIC is a function of process, voltage, and temperature (PVT) and assumes that the entire device is impacted by a single PVT condition (e.g., worst case). In today's submicron process, however, it is imperative to realize that a more complicated PVT needs to be considered.

The PVT profile complexity (or OCV) tends to degrade the chip performance. Usually, the impact of these types of variation on chip timing cannot be analyzed by standard timing checks using single PVT conditions such as worst-case timing scenarios for setup checks or best-case timing scenarios for hold checks.

In performing OCV analysis, one must compare a computed pair of timing arc delays. For setup checks, the worst PVT condition is used for clock and data paths (launch path), and the best PVT condition is used for reference clock (capture path). Similarly, for hold timing checks, the best PVT condition is applied to clock and

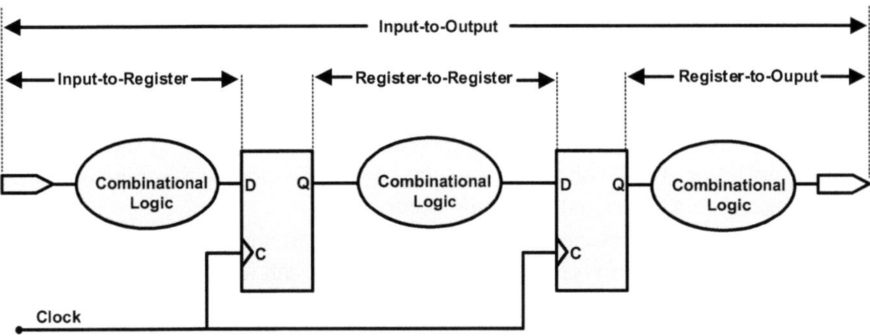

Fig. 5.9 Typical timing paths

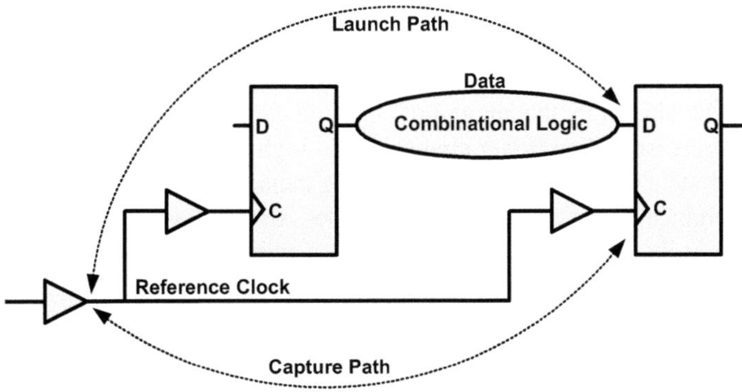

Fig. 5.10 Launch and capture path

data paths, and the worst PVT condition is applied to the reference clock. An example of launch and capture paths is shown in Fig. 5.10.

Another timing analysis consideration is the effect of resistance and capacitance coupling. High coupling resistance and capacitance in deep submicron process result in both noise and additional delays that greatly affect the functionality and performance of any ASIC design.

Because the number of silicon failures caused by undetected and unresolved resistive-capacitive coupling violations is rising dramatically, it is important to minimize their occurrence. This can be accomplished by various methods such as wire spacing and shielding during the physical design stage and analyzing their impact on signal integrity during design timing closure.

The most common effects of interconnect coupling capacitance are induced by noise and delay. The crosstalk-induced noise occurs when signals in adjacent wires transition between logic values and capacitive coupling between wires causes a charge transfer [3].

Depending on the rate of change of the signal and the amount of coupling capacitance, there can be a significant possibility of noise injection between the aggressor and victim wires. For a capacitively coupled system, the aggressor is the wire that has the most signal activity, and the victim is the quiet one, as illustrated in Fig. 5.11.

In the very simplistic illustration presented in Fig. 5.11, a fast transition from the driver of the aggressor wire induces a glitch or noise at the input of the receiver of the victim wire regardless of its voltage level (logic 0 or logic 1). If the voltage of ensuing noises or glitches on the victim wire crosses the input of the receiver threshold voltage, then a functional error may occur.

These types of errors can propagate through combinational logic and subsequently alter the state of registers in the design. This can lead to design functional failure.

Apart from noise injection due to crosstalk capacitance, this capacitance also has a serious impact on the adjacent wire delays. This situation becomes even more complex when signals simultaneously switch on both aggressor and victim interconnects.

For illustration, consider two parallel interconnect lines driven by logic gates shown in Fig. 5.11. Depending on the activity of these lines, their effective capacitance may be altered by the amount of coupling capacitance.

If the aggressor's line driver changes from high to low and the victim line is in the ground state (logic 0), then the coupling capacitance may cause the effective load capacitance on the aggressor line to be less than $(C_a + C_x)$ or $(C_v + C_x)$. On the other hand, if the victim's line driver is in the power supply state (logic 1), then the effective capacitance of the aggressor line may exceed $(C_a + C_x)$ or $(C_v + C_x)$.

The same effect can also be observed when the transition changes in the opposite direction for both aggressor and victim lines. In this situation, the effective load capacitance deviates from $(C_a + 2C_x)$ for the aggressor line and $(C_v + 2C_x)$ for the victim line.

The situations described here are somewhat simplistic. In practice, many topological possibilities and conditions exist that could contribute to the delay uncertainty for any given ASIC design. Therefore, the availability of an efficient and

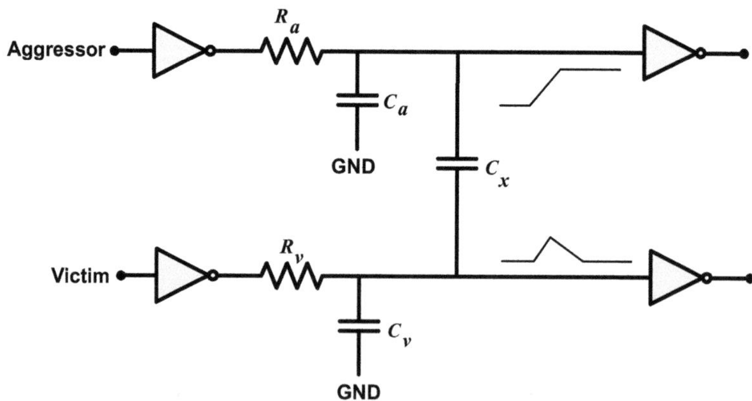

Fig. 5.11 Crosstalk-induced noise between aggressor and victim

accurate timing engine to estimate the delay of a coupled system is a crucial element for crosstalk avoidance during physical design timing verification.

There are several design techniques that can be applied to minimize coupling effects. The most effective one is to increase the spacing between neighboring wires of the same metallization as described in the previous chapter. While this technique may be effective for selective interconnects such as clock signals, it is not economically efficient for global application due to its wiring area impact [4].

Another method that is used to minimize, or even eliminate, the delay unpredictability in a capacitive coupled system is to make sure that the aggressor and the victim driver strength and corresponding capacitive loading are the same. This is an effective method under an in-phase signal transition. In the case of out-of-phase signal transition, to reduce propagation delay due to crosstalk, the engineer needs to adjust the drive strength of the aggressor and/or the victim lines such that there is no transfer signal delay from one path to another path through the effective capacitance for various data paths.

STA is an integral part of any ASIC design flow. However, for accurately studying critical paths and crosstalk-induced noise and delays, the use of highly sophisticated timing engines that can account for high-order effects using transistor-level dynamic analysis is required.

Apart from static timing, one should not underestimate the advantages that dynamic simulation can offer. For a comprehensive timing verification and validation methodology, one should include both static and dynamic timing analysis in their ASIC design flow.

Using dynamic timing simulation as shown in Fig. 5.12, analysis can be advantageous in a sense that one can analyze the design timing requirement without breaking any timing loop; it is not possible to study this under static conditions. Moreover,

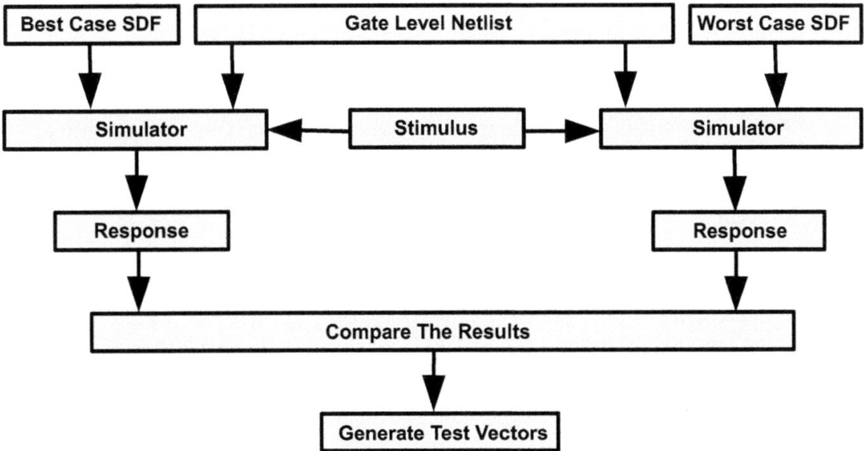

Fig. 5.12 Dynamic simulation flow

the result of dynamic simulation can be used during the performance testing (or at speed testing) of ASIC devices.

5.3 Physical Verification

The primary objective of physical verification is to check the ASIC layout against process rules provided by semiconductor foundries to ensure that it can be manufactured correctly. However, for advanced process nodes (i.e., 7 nm and below), there are additional requirements for checking an ASIC physical design such as Design Rule Checking (DRC) and Product Reliability Checking (PRC).

Both DRC and PRC are process of verifying that a physical design of an ASIC design meets the design rules and specifications of a particular technology or foundry. However, they have different purposes and scopes. DRC is a process of checking the physical layout data against fabrication-specific rules specified by the foundry to ensure successful fabrication. DRC checks for basic and common types of rules, such as minimum width, minimum spacing, minimum area, wide metal jog, misaligned via wire, etc. DRC is an essential part of the physical design flow and ensures the design meets manufacturing requirements and will not result in a chip failure.

PRC is a process of checking the physical design for more advanced and complex rules that are related to the electrical behavior and performance for a given ASIC design. PRC can be divided into two types: Physical Rule Checking and Physical Reliability Checking. Physical Rule Checking checks for rules that are related to the physical integrity and functionality of the ASIC device, such as antenna checks, density checks, lithography checks, etc. Physical Reliability Checking checks for rules that are related to the reliability and robustness of the IC against various physical effects, such as electromigration, self-heating, thermal stress, electrostatic discharge, Latch-up, etc. PRC is an important and necessary step to ensure the quality and performance of an ASIC.

The process of physical verification is highly automated and will continue to become more automated in the future. As ASIC designs and their fabrication process become more complex, physical verification software advances. Dramatically decreased device sizes, with respect to the number of transistors and routing layers involved in the net result for physical verification of larger layout databases, need to be verified against rules that are more complex. Verifying this amount of data puts strain on the software and computer infrastructure. Therefore, to improve physical verification efficiency and debugging procedures, one recommendation is to make sure that the ASIC is physically designed and implemented by employing a correct-by-construction methodology. Several checks are performed during physical verification. Mainly these checks are Layout Versus Schematic (LVS), DRC, and Electrical Rule Check (ERC).

LVS verification examines two electrical circuits to see if they are equivalent with respect to their connectivity and total transistor count. One electrical circuit corresponds to transistor-level schematics, or netlist (reference), and another electrical circuit is the result of the extracted netlist from the physical database. Through the verification process, if the extracted netlist (layout) is equivalent to the transistor-level netlist, they should function identically.

LVS verification is fast becoming challenging for today's ASIC designs that consist of millions of transistors and enormous amounts of specialized data and deep submicron effects. Although current LVS verification software efficiently provides layout-based error reports, error diagnosis remains a concern.

One of the primary issues with LVS verification is the repeated iteration of design checking needed to find and remove comparison errors between the layout extracted and the transistor-level netlist. The cycles involved in the LVS verification consist of data output (GDS) from the physical database, transistor-level generated netlist, the actual LVS run itself, error diagnosis, and error correction. Thus, one of the objectives during LVS is to keep the length of time needed to complete the verification cycle as short as possible.

Currently, there are two ways to improve the LVS verification cycle time. The first consideration must be given to the capacity and performance of the machine infrastructure along with the type of physical verification software. Physical verification software must execute fast and provide accurate results that can be easily traced in the case of error. The second is to use hierarchical verification features rather than flat comparison.

Using hierarchical verification features are very important, not only to minimize the amount of data to check but also to identify the errors through the usage of hierarchical cells and black box techniques.

It should be pointed out that although the hierarchical verification method is far superior to that of the flat approach, LVS software that incorporates both hierarchical and flat comparison is far better than verification software that is either strictly hierarchical or flat.

By recognizing the components that simply match between the netlist and layout (such as standard cell libraries), memory blocks and IP elements can be compared in a hierarchical manner, while allowing other design elements such as analog blocks and macro cells to maintain a flat representation. In this way, debugging performance is drastically enhanced.

Finally, but most importantly, it is recommended to begin the verification process during an early stage of the design to ensure that the physical database is correct. Under the assumption that most of today's physical design tools are capable of producing error-free place-and-routed designs, the most common sources of physical errors arise during the floorplanning stage and most often are related to power and ground connections. Power and ground shorts, and/or opens, impact device (transistor) recognition during LVS verification, which then leads to an extremely long execution time.

DRC is considered a prescription for preparing photo-masks that are used in the fabrication of the ASIC design. The major objective of the layout or DRC is to obtain optimal circuit yield without design reliability losses. The more conservative the design rules are, the more likely it leads to correctly manufactured ASIC designs.

However, the more aggressive the design rules are, the greater the probability of yield losses.

DRC software uses so-called DRC decks during the verification process. Naturally, DRC decks should contain all manufacturing design rules. As semiconductor manufacturing becomes more complex, so do the DRC decks. These complex DRC decks must be written or composed in an efficient manner. If the layout rule checks are not coded in an optimized way in a DRC deck, they tend to require more time (machine run time) or resources (machine memory) to complete verification.

Reducing DRC verification run time so that the rule deck is efficient from a coding point of view is desired. Like LVS verification, one also needs to consider utilizing the hierarchical process.

Hierarchical processing enables the tools to recognize multiple occurring instances throughout the physical database and checks each instance one at a time, rather than checking all instances concurrently. Another advantage of hierarchical DRC verification is that the amount of data that needs to be checked is reduced together with the machine process and memory requirements.

Another consideration is to use a comprehensive DRC deck during physical verification. If the physical verification is not performed using a comprehensive DRC deck, the result can be low yield or no yield at all. For a DRC verification to be considered comprehensive, one must make sure the DRC deck checks all the rules correctly and accurately and is able to properly identify and address yield-limiting issues. The most common yield-limiting issues are:

- Charge accumulation due to antenna effects
- Inadequate planarity for multiple layers that are required by Chemical Mechanical Polishing (CMP)
- Metal line mechanical stress
- Electrostatic Discharge (ESD) and latch-up

Regarding the antenna problem during the metallization, there are two methods of interest: one is ratio calculation, and another is the wire-charged accumulation. During the DRC verification, ratio calculation in conjunction with wire-charged accumulation is used. In the ratio calculation method, the following ratios can be calculated:

- Wire length to the connecting gate width
- Wire perimeter to the connecting gate area
- Wire area to the connecting gate area

Like the ratio calculation method, for wire-charged accumulation, assuming N is the metal layer currently to be etched, the following methods can be considered:

- Layer N connecting to the gate
- Layer N plus all layers below forming a path to the gate
- Layer N plus all layers below layer N

Another aspect of DRC allows one to check for DFM such as contact/via overlaps and end-of-line enclosures. The DFM rules are considered optional and are provided by the silicon manufacturer. From a yield perspective, it is beneficial to check the ASIC physical design for DFM rule violations and then correct the errors as much as possible without influencing the overall die area.

ERC verification is intended to verify an ASIC design electrically. In comparison with LVS that verifies the equivalence between the reference and extracted netlists, ERC checks for electrical errors such as open input pins or conflicting outputs. A design can pass LVS verification but may fail to pass ERC checks. For example, if there is an unused input in the reference netlist, then the extracted (routed) netlist will also contain the same topology. In this case, the result of the LVS process will be correct by matching both circuits, whereas the same circuits will cause an error during ERC verification (since a floating gate can lead to excess current leakage). In the past, ERC verification was used to check the quality of manually captured schematics. Since manual schematic capturing is no longer used for digital designs, ERC verification is mainly performed on the physical data. The ERC verification process is a custom verification rather than a generic verification. The user can define many electrical rules for the purpose of verification. These rules can be as simple as checking for floating wires or more complex, such as identifying the number of PWELL- or NWELL-to-substrate contacts, latch-up, and ESD [5].

5.4 Summary

In this chapter, the ASIC functional, timing, and physical verification processes are discussed. In the functional verification section, various methods such as simulation-based and rule-based verification style are briefly explored. In addition, for the rule-based method, it covered the basic concepts of assertion and formal methods.

The timing verification section outlined the basics of wire delay calculation rather than its extensive mathematical derivation. Also overviewed were the effective and coupling capacitances and their relationship with timing verification. In addition, methods for generating false paths to accurately verify design timing requirements were discussed. Noise effects on timing were explained, and the reader was provided some basic methods for preventing their impacts on the timing of the design.

In the physical verification section, the basic steps that are required to verify an ASIC design physically, such as LVS, DRC, PRC, and ERC, were provided. Verification steps integrate all phases of the design process, and, in today's world, ASIC design verification is severely constrained by human and computer resources. In addition, it should be noted that as the number of verification cycles required to verify an ASIC design grows exponentially (as a function of gate-level complexity), then the solution to ASIC verification productivity problems is to deploy significantly more effective verification methods and tools. This requires a thorough understanding of the system and verification goals, process, and technology.

References

1. Khosrow Golshan, *Physical Design Essentials, an ASIC Design Implementation Perspective*, Springer Business Media, 2007
2. Khosrow Golshan, *The Art of Timing Closure, Advanced ASIC Design Implementation*, Springer Nature, Switzerland AG, 2020
3. A. Vittal and M. Marek-Sadowsks, "Crosstalk Reduction for VLSI", *IEEE Transaction on CAD*, Vol. 16, No.3, March 1997
4. Kevin T. Tang and Eby G. Friedman, "Delay and noise estimation of CMOS logic gates driving coupled resistive-capacitive interconnections", *INTEGRATION, the VLSI Journal*, Vol. 20, 2000
5. Sanjay Dabral and Timothy J. Maloney, *Basic ESD and I/O Design*, John Wiley & Sons, Inc. Publishing Company, 1998

Chapter 6
ASIC Testing

> *The only reason for time is so that everything doesn't happen at once.*
>
> –Albert Einstein

The rapid growth in the size of ASIC devices, the complexity of designs, and tighter design geometries are making test methods challenging. Device-level testing consumes a great deal of time and requires sophisticated programs necessitated by heightened reliability and performance standards required by many of today's products and complex Automated Test Equipment (ATE). This results in increasing costs and development time. There are many innovative techniques proposed to address this problem. One of the most profound techniques is to incorporate DFT.

The concept of DFT of an ASIC design is primarily related to both the controllability and observability of its internal nodes. In general, the circuitry of an ASIC needs to be designed in such a way that the internal nodes of the circuit can be controlled from the primary inputs and observed through the primary outputs. In contrast with board-level testing, where it is often quite easy to probe circuit nodes, with device-level testing, the ASIC must be fully testable from its package pins or I/O pads.

Unfortunately, there is no specific formula that can be applied to provide the answers to all DFT issues, as each design will be subject to different circumstances. However, by discussing the various conditions involved, the ASIC and physical designers must decide on the optimal or ideal test philosophy applicable to that design. The optimal or ideal test has a precise definition which can be applied with ease and maximum effectiveness (an ability of test programs unambiguously to distinguish faulty devices from those that perform properly over the range of specified operating conditions).

Since a fundamental requirement of any manufacturing process is the ability to verify that the finished product will successfully operate its intended function, any ASIC device should be subjected to sufficiently rigorous testing to demonstrate that the product's design requirements will meet various stages of the design. These testing stages are:

K. Golshan, *ASIC Design Implementation Process*,
https://doi.org/10.1007/978-3-031-58653-8_6

- Circuit and physical design
- Device processing (wafer-level testing)
- Device packaging (package-level testing)
- Device mounted to the circuit board (board-level testing)
- Circuit board insulated in the system (system-level testing)
- Product delivery (field-level testing)

It is important to note that the cost to detect and fix a defect increases exponentially as the product moves from early design to final product delivery. For example, if it would cost N units to fix a problem during the design phase, the same problem might cost 1000 N (this is an arbitrary scale and depends on the various organization business models) to be fixed during field test. Figure 6.1 illustrates test costs at different ASIC testing stages.

The reason for this increase in the cost to fix design defects during field testing is because the only requirement to fix the defects during the design phase is to have a comprehensive test program, whereas fixing the faults in field testing relies on more complex and costly business and engineering structures. Therefore, it is imperative to fix any problems during the early stage of the ASIC design to ensure system function reliability in the field.

In order to ensure manufacturability, the following are the most common tests performed:

- Functional test
- Scan test
- Boundary scan test
- Fault detection
- Parametric test
- Current and very low-level voltage test
- Built-In-Self-Test
- Parallel module test

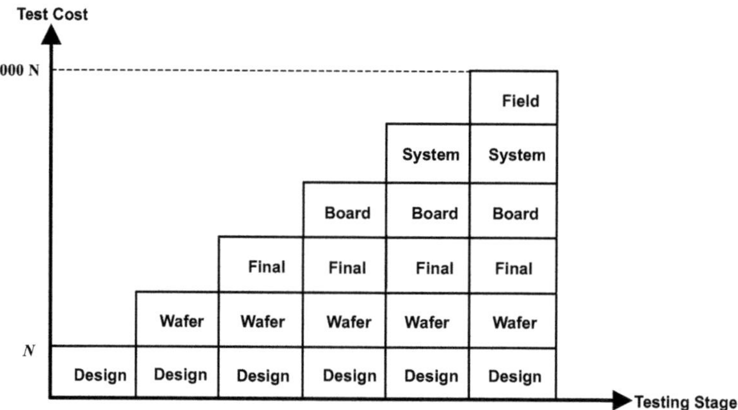

Fig. 6.1 Test cost at different testing stages

6.1 Functional Test

The primary reason for functional, or at-speed, testing is to determine that the circuit logic is functioning correctly. In addition, it is necessary to establish that both AC and DC performance criteria of the design will be met over the full range of electrical and environmental conditions.

One of the prerequisites of functional testing is to create a set of comprehensive stimuli or test vectors that exercise the logic. These test vectors can be applied at the behavioral, or RTL, level or at the structural gate level. The advantage of simulating logic circuits at the behavioral level is shorter execution time, as no specific timing associated with the design is required. However, because of the lack of specific timing information, the result of behavioral testing cannot be imported to any test equipment for future testing.

For creating a set of test vectors that can be imported to ATE, the final netlist from the physical design or place-and-route tools, along with its parasitic information, is used. During test pattern generation, the relationship and sequence of input stimuli applied to the ASIC design to test the device performance and functionality are different from those that are used for circuit simulation used to verify the logic design. These differences are mainly due to the system requirement of ATE; thus, test vectors need to be simulated consistently with targeted ATE capabilities and their margin constraints.

Margin constraint refers to a value that is greater than the actual ATE input placement accuracy and is intended to prevent process-induced yield losses due to input margins. Margin constraint is used during input margin testing on a single process lot that cannot represent the full min-delay and max-delay process window. To overcome this limitation, input skew margins over min-delay and max-delay conditions, simulation, and/or STA can be used. Some of the fundamental ATE capabilities are related to their:

- Operating frequency
- On/off clock width
- Number of signal pins
- Loading transistors such as CMOS, TTL, etc.
- Input skew and placement accuracy
- Output strobe timing and stability

Generating a set of test programs that will completely cover the functional blocks in the design is not an easy task, especially if the circuitry was not designed with DFT in mind. As mentioned earlier, one aim in the design of ASIC devices is to design the circuit in such a way that its design function allows its internal nodes to be controllable and observable from primary inputs and outputs. This will increase the ability to determine the nature of any fault and can be particularly useful during the prototyping debugging stage.

In the case of complex ASIC designs, incorporating DFT features, such as proper circuit initialization, asynchronous design avoidance, and use of multiplexers, increases controllability and observability of the device and leads to improved testability. Proper circuit initialization ensures predictable and repeatable operation of each circuit node under test conditions regardless of their power-up condition. Initialization conditioning may not be as important for combinational logic; however, it is important when dealing with sequential logic (especially if sequential circuits require a complex test pattern or test programs). In that case, one may need to consider inclusion of a master reset from a primary input pad to enable all initialization of all registers in the design. Using reset lines can also be advantageous during logic simulation, particularly those that include unknown states.

During logic simulation to generate test vectors for device testing, one should not initialize unknown internal nodes by forcing logic to "0" or "1," as a failure may occur during actual device testing.

Asynchronous design is another example that may require additional consideration regarding DFT. Although synchronous design techniques are always preferred, some occasions may require asynchronous circuit design. Asynchronous design, by its nature, is not only difficult to control but also may lead to other timing problems, such as races or glitches, in decoding logic. When dealing with asynchronous circuitry, extreme care must be practiced during logic and physical design in order to avoid any timing hazards. Use of multiplexers may increase both controllability and observability during the test phase. An inaccessible circuit has low circuit observability and, therefore, is very difficult to test. One often-used method of testing is to connect non-observable nodes directly to a primary output using a multiplexer, as shown in Fig. 6.2.

In this example, the non-observable D input of the second Flip-Flop is connected to the A input of the multiplexer, allowing the D input to become an observable node.

Multiplexers may also be added to the circuit for improved controllability. Designing for controllability often reduces the number of test patterns required and should be considered when long dividers or counters are present in the design, or for initialization purposes, as shown in Fig. 6.3. In this case, inserting a multiplexer at the D input of the Flip-Flop allows the circuit to be initialized from a primary input.

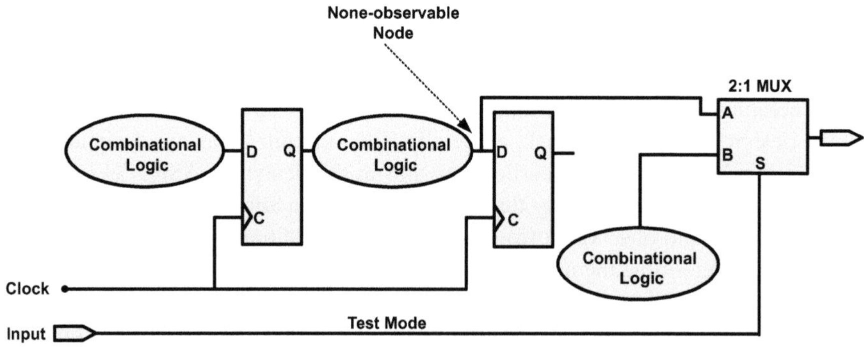

Fig. 6.2 Non-observable node

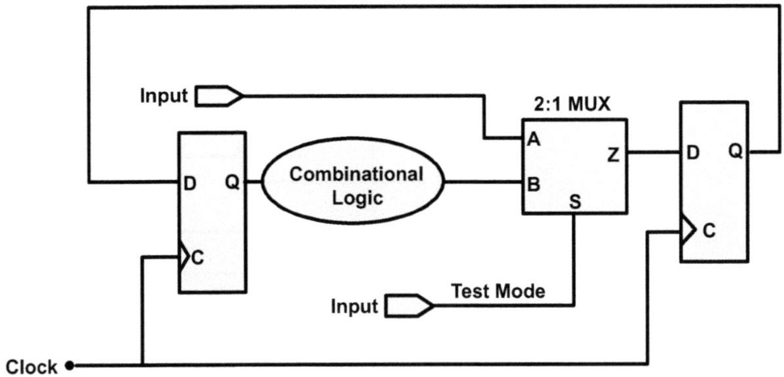

Fig. 6.3 Multiplexer initialization usage

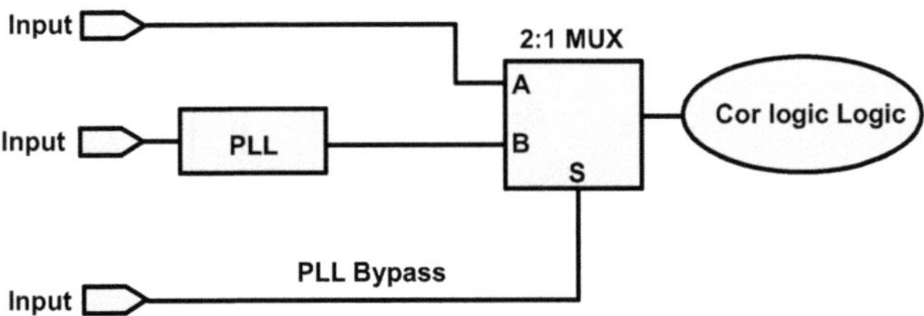

Fig. 6.4 Multiplexer usage as PLL bypass mode

In addition, multiplexers can be used as bypassing elements when a design contains an on-chip clock oscillator or PLL that cannot be controlled externally, as shown in Fig. 6.4. This is important if it is not possible to synchronize the ATE to the DUT during test.

6.2   Scan Test

Scan test is classified as a test for manufacturing and is used to detect stuck-at faults (stuck-at-zero and stuck-at-one) that may occur during the fabrication process. The main objective of scan test implementation is to achieve total, or near total, controllability and observability in sequential circuits. In this method, ASIC design uses scan Flip-Flops, latches, or both, to operate in either parallel (normal) mode or serial (test) mode.

The scan Flip-Flop is designed to provide scan-testable features in edge-sensitive ASIC design. These scan Flip-Flops can be used in place of normal Flip-Flops when designing registers, counters, state machines, etc.

A scan Flip-Flop consists of a basic Flip-Flop and a 2–1 multiplexer as shown in Fig. 6.5, with SI as scan input, SE as scan enable, DI as the normal data port, CK as input clock, and DO as data output.

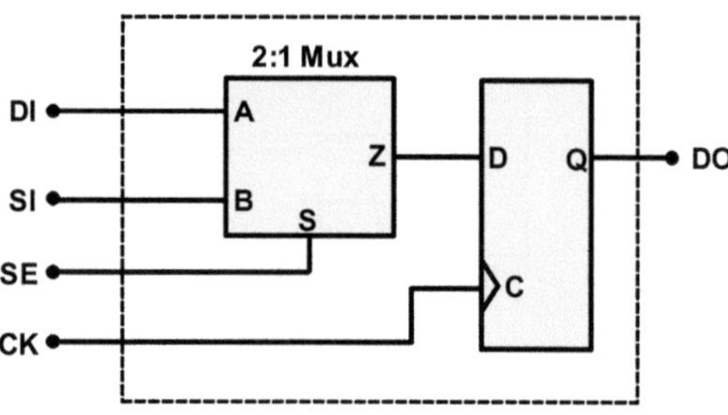

Fig. 6.5 Basic scan
Flip-Flop configuration

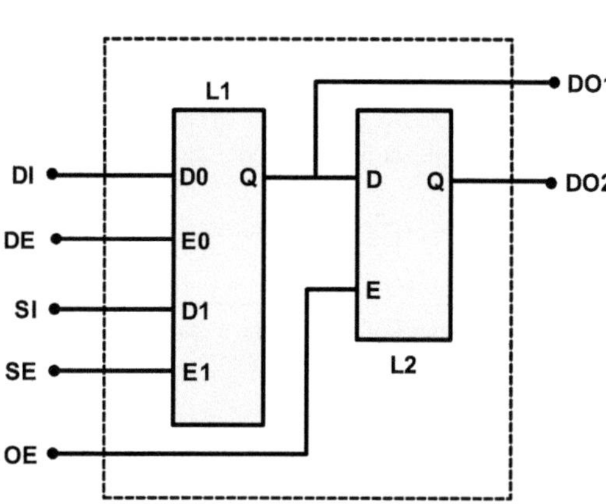

Fig. 6.6 Basic scan latch
configuration

In the normal or functional mode, the scan Flip-Flops are configured for parallel
operation (or capture mode). In test (or scan) mode, the scan Flip-Flops are loaded
(controlled) by serially clocking in the desired data (or shift mode).

The scan latch is designed to provide scan-testable features in level-sensitive
ASIC design. These scan latches can be used in place of normal latches when
designing registers, counters, state machines, etc. Level Sensitive Scan Design
(LSSD) addresses the problems associated with implementing the scan Flip-Flop
design [1].

LSSD improves the scan design concept by making the scan cells level-sensitive.
In a level-sensitive system, the steady-state response to input changes is independent of circuit and wire delays within the system. One of the challenges using this
technique is that LSSD imposes constraints on circuit excitation, particularly in
handling clocked circuitry.

Scan latches consist of a basic dual-port latch (L1) and a basic single-port latch
(L2) as shown in Fig. 6.6.

Multiple input enable-ports (DE, SE, and OE) control the latch's outputs (DO1
AND DO2). In normal operating (or capture) mode, enable DE latches data into L1
from the data input (DI), while scan enable (SE) is inactive. In the scan (or testing)
mode, scan enable (SE) latches in the scan data (SI), while data enable (DE) is inactive. Output enable (OE) transfers this data from L1 to L2. In this mode of operation, the output data can be taken from either DO1 of L1 or DO2 of L2.

In today's complex ASIC designs, scan testing is becoming an integral part of the design process to achieve a higher level of fault detection after manufacturing. In the physical design flow, all functional design constraints must be met in the presence of scan logic. During placement and CTS phases, the physical information (or standard cells' physical locations) of the design is used to minimize routing congestion due to scan chain connectivity as well as minimize functional and scan clock skews.

For minimizing routing congestion during the placement phase, scan chain reordering is performed. It is important to have physical design tools that can optimize the reordered chain for both timing and power.

Consideration of clock distribution and balancing to minimize clock skew in both functional and scan mode is the next step. This may present some challenges as most of today's ASIC designs consist of multilevel clock gating elements, clock dividing, mode switching circuits, and scan clock. Most of today's CTS tools can generate quality clock trees. However, when clocking logic becomes more complex because of intensive clock gating logic and/or multiple mode-switching logic, these tools may not guarantee construction of high-quality clock distribution regarding skew budgeting and buffer-tree balancing. Thus, it is imperative that physical design tools and their CTS algorithms perform multimode clock synthesis and be able to meet both functional and scan timing requirements simultaneously.

Today's complex ASIC design contains millions of Flip-Flops, and a single scan chain can be costly during scan test. In order to reduce the scan test time, scan compression is needed for today's ASIC designs because it helps to reduce the test data volume and test application time, which are important factors for reducing the test cost and improving the test quality. Scan compression also enables shorter scan chains, which can reduce the scan-shift power. Scan compression is a technique that allows testing of ASICs with fewer scan I/Os and shorter scan chains as shown in Fig. 6.7.

Scan compression works by inserting compression logic in the scan path between the scan I/O and the internal scan chains. The compression logic consists of a decompressor on the scan input side and a compressor on the scan output side.

The compressor fans out the scan inputs to the internal scan chain inputs, and the decompressor compares (e.g., XORs) the internal scan chain outputs to connect them to the design scan outputs. This way, the scan data can be compressed and decompressed during the test, reducing the test time and data volume.

Scan compression is used in nearly every design, especially for complex and large-scale designs that have new fault models and additional test patterns. Scan compression inserts compression logic in the scan path between the scan input, scan output, and the internal chains, which breaks the link between them and allows for more internal scan chains with shorter lengths. There are different techniques and architectures for scan compression, such as decompressors and compactors, which can achieve different compression ratios and trade-offs.

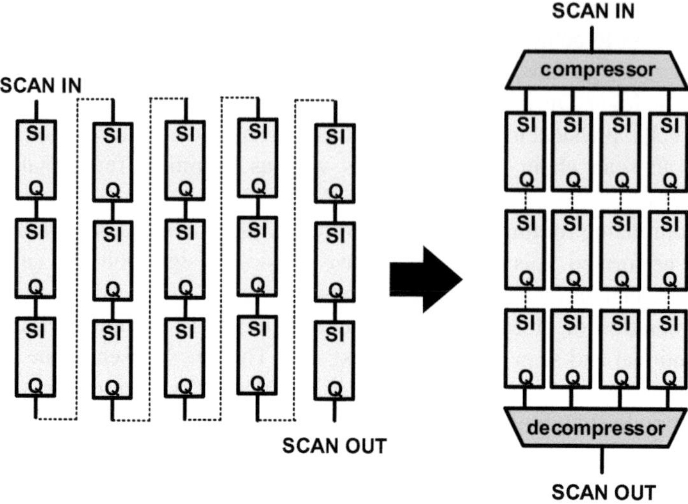

Fig. 6.7 Scan chain compression example

6.3   Boundary Scan Test

Boundary scan design extends the scan design methodology for testing primary inputs and outputs of an ASIC design. This technique is suited for solving problems resulting from test equipment costs and testing difficulties with surface-mount technology such as automated probing.

This technique allows designers to place boundary scan cells next to each I/O pad in the design for improved controllability and observability within and between the chips.

At the chip level, boundary scan cells have other terminals through which they can be connected to each other. They then form a shift register path around the periphery of the ASIC, thus making it possible to test the package pins. Figure 6.8 shows a basic boundary scan cell consisting of two 2:1 multiplexer and one Flip-Flop.

During normal operation, data passes between the ASIC I/O pins and internal core logic through data-in and data-out ports as if the boundary scan cells were transparent. When the ASIC is in test or boundary scan mode, however, the data passes through the boundary scan cells' shift register paths between serial-in and serial-out ports. By loading data into the boundary scan cells, the boundary scan cells can prevent the data flow to or from the ASIC I/O pads, so that the ASIC can be tested for either its internal logic or for external chip-to-chip connections. Using the concept of shift register paths, arbitrary data can be loaded into the boundary scan cells, and data can be observed from those boundary scan cells. The advantage of boundary scan testing is that it simplifies the test pattern generation required to create test vectors for an ASIC design. To use the boundary scan cells, additional

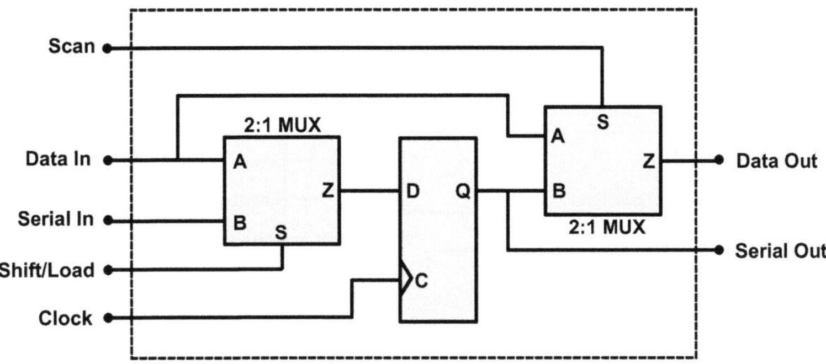

Fig. 6.8 Basic boundary scan cell

I/O pads are required, as well as some additional logic on the ASIC design to control the boundary scan chain for device testing.

IEEE has addressed these boundary scan requirements by setting a standard: ANSI/IEEE Standard 1149.1 IEEE Standard Test Access Port and Boundary-Scan Architecture [2]. This standard defines the I/O pins (Test Access Port or TAP), control logic, and TAP controller (state machine) that generate the control signals required to operate the ATE for boundary scan testing of ASIC devices.

ANSI/IEEE Standard 114.1 gives details on expansion of the tool set for chip test designs. It also defines a method for communicating test instructions and data from an external processor to the ASIC. Based on these requirements, the ASIC I/O pads can be configured to meet the IEEE Standard 1149.1 specifications. Figure 6.9 illustrates basic boundary scan chain configurations and basic TAP controller connections. In Fig. 6.8, the additional pins required for an ASIC design to implement boundary scan testing features are as follows:

- The Test Data In, or TDI, that connects to the TAP controller and the first boundary cell serial-in port in the chain
- The boundary scan Test Clock, or TCK, that connects to all boundary scan clock ports in parallel
- The control signal TMS connects to the TAP controller for generation of proper sequencing of test signals and is also connected to the scan and shift/load ports of all boundary scan ports
- The Test Data Out, or TDO, that connects to the last boundary scan cell serial-out port in the chain

6.4 Fault Detection

The purpose of fault detection testing is to detect defects that occur during ASIC manufacturing. These defects can be categorized either as hard or soft defects. Hard defects are related to shorted (or bridged) interconnects, gate oxide defects, and

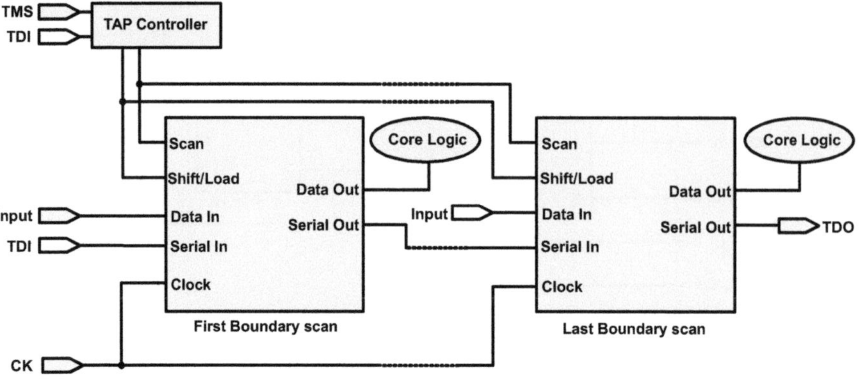

Fig. 6.9 Boundary scan chain configurations

open interconnects. Detecting hard defects is based on changes measured as logic-level voltages such as stuck-at-one (SA1) or stuck-at-zero (SA0). Although SA1 and SA0 fault testing has become standard, it has proven insufficient for thoroughly detecting all hard defects, as it only detects nodal defects.

The model for SA1 and SA0 fault testing (e.g., scan) uses zero-delay timing logic-level representation of a circuit at the primitive cell, or gate, level. Using this type of fault model, the flow requires running a set of test patterns that drive each circuit node high and low and any SA1 and SA0 signal to the output pins. These output signals are then compared to an expected Good Known Signal (GKS) to detect defects. In the case of scan testing, these test patterns are generated automatically using specific software known as Automatic Test Pattern Generation or ATPG.

ATPG software uses a fault model for the circuit under test and creates a list of all faults or fault dictionary. This fault dictionary lists all the fault information about each node in the design. Often the fault dictionary will be in condensed format, since redundant SA1 and SA0 faults for a given node in the circuit can be removed.

The information listed in the fault dictionary indicates whether the test pattern generated by ATPG tools has detected SA1 and SA0 at each node in the design. In the process of generating ATPG test vectors, at least 97%, or better, test coverage of SA0 and SA1 is an industry-common practice for detecting nodal defects. There are three distinct operations that take place when using ATPG to generate scan test patterns: register path conditioning (or scan mode), data capturing (or normal mode), and data shifting (or scan mode).

Register path conditioning is used to make sure all Flip-Flops in a given scan path are operating properly and are set to a known value (i.e., initialization state) before the actual scan test operation. During this mode, the circuit is brought up to scan mode, and then a series of 1 and 0 data are serially shifted through all registers from the primary scan input until they reach the scan primary output (i.e., the scan chain is fully loaded) of the chip. This operation requires one scan clock for each Flip-Flop element in the scan chain.

In capture mode, the circuit is brought up to normal mode, and then the data from the combinational logic elements registers is clocked into scan Flip-Flops using scan clock while applying data at primary inputs of the chip. In the last, and third, part of the scan operation, or scan mode, the circuit is brought up to scan mode, and scan clock is used to unload the scan chain through the scan output. During this operation, captured data shifts out of the scan chain, and the new data is loaded from the next test pattern into the scan chain by simultaneously asserting and reasserting the scan-enable signal on the falling edge of the scan clock to ease timing associated with hold time constraints.

Although one can achieve reasonable stuck-at fault coverage using scan techniques with the aid of ATPG, this method does not necessarily detect all possible CMOS defects such as voltage weakening or delay faults. These types of faults are commonly referred to as soft detections and are mainly timing related. At-speed functional testing is the most often used test approach in detection of soft defects. At-speed functional test patterns provide an excellent defect detection mechanism for any design (if prepared properly).

In contrast with the hard defect model, the soft defect (slow-to-rise and slow-to-fall) fault model uses actual delay models for each element, or gate, in the logic-level representation, or structural netlist, to create a fault dictionary. The fault dictionary is used during the test program development to determine the quality of the test vectors (i.e., test coverage).

One of the advantages of at-speed test vectors is the byproduct of design development and functional verification (development time). However, the disadvantage of this type of test vector is the difficulty in measuring fault coverage. The problem of measuring the fault coverage of functional, or at-speed, test vectors has led to the development of software tools that perform testability analysis, otherwise known as fault simulators.

Fault simulation allows testability (or controllability and observability) of the ASIC device to be analyzed during the design phase by identifying those areas of the design that have poor testability, thus allowing the designer to act appropriately to improve the design testability.

6.5 Parametric Test

In addition to at-speed, or functional, testing, it is important to ascertain whether the ASIC design will operate within its specified AC timing or DC interface parameters. These types of test programs are known as parametric and test all critical AC and DC parameters to ensure that the device will operate successfully over the recommended supply voltage and ambient temperature ranges. AC performance can be obtained from at-speed, or functional, testing of the ASIC design if the ATE capabilities support this. AC parametric test vectors verify timing requirements such as propagation delay (or path delay) and maximum operating frequency of the design. Other timing-related parameters, such as primary input setup and hold times or propagation delays of the primary outputs, may need to be tested and measured separately. This can be accomplished by choosing an appropriate location in the test vector and programming the ATE to measure the relevant AC parameters.

As design complexity and gate counts increase with process technology, AC testing for controllability and observability of the critical timing path can present a limitation. Therefore, it is important to be able to expose critical paths automatically (like scan ATPG) in generating AC parametric test programs.

DC testing is intended to test adherence to primary input and output voltage and current requirements. Testing a single DC value often requires a set of test programs—some to set up output pins and others to create the signal value being measured. Thus, the fundamental requirement for DC testing is to be able to measure all DC parameters on each primary input and output as well as power supply currents for a given ASIC design.

It is desirable to have an ATE that is capable of logically operating the device at maximum operating frequency, measuring DC parameters, while continuously varying operating conditions such as voltage and temperature. However, in practice this is impossible; therefore, one may select a few operational conditions, such as minimum and maximum voltage and temperature.

One of the DC test requirements is to set each buffer in the I/O pad to all available states. It is important to note that the DC parametric testing is not performed for missing toggle states, but rather it measures the electrical property of each I/O pad in the design. DC specified parameters can only be carried out with a detailed knowledge of the I/O pad circuits and characteristics of the process used to fabricate the ASIC design.

6.6 Current and Very Low-Level Voltage Test

Current (or I_{DDQ}) and Very Low-level Voltage (VLV) testing can detect certain defects and degradations that cause parametric failures (or soft defects) not detected by functional at-speed test programs. These types of soft defects can lead to timing failures that make a circuit fail to operate at the designed speed while being functional at a lower speed [3].

I_{DDQ} testing and VLV offer alternative testing strategies that can detect many additional defects and reduce defect levels by an order of magnitude. Conventional test programs are the best for detecting some defects such as interconnect shorts and opens, while I_{DDQ} and VLV testing is best for detecting the other defects such as via defects, stress voids, increase in interconnect delay due to electromigration damage, threshold voltage shift, gate oxide shorts, and tunneling effect.

I_{DDQ} testing measures current faults and is mainly used for CMOS processing due to its low quiescent current requirement. In general, current faults (or excessive current) are observed when the low quiescent current of a CMOS device abruptly increases due to an undesired conducting path (or short) between power and ground.

The current fault model for CMOS devices is a transistor-level model of a circuit with each transistor (PMOS and NMOS) operating with proper quiescent current once they have completed their logical transitions. I_{DDQ} testing requires executing a set of test patterns on the ATE that drive each circuit node to a high and low value. After applying each test pattern, the current through the power supply pads (or VDD) is measured after allowing the logic transitions to complete. These currents are then measured against limits set by normal static current of the ASIC to detect faults.

The advantage of these types of tests is that they are much less pattern-sensitive in comparison with the scan or functional tests as faults are visible through power pads and do not have to be logically propagated out of a device. Therefore, it is not necessary to create a well-known output test program. I_{DDQ} tests are considered to be more suited for detecting faults such as bridged interconnects, gate oxide shorts, and tunneling opens (a very thin layer of oxide located in via or contact causing transistor gates to be disconnected) that cannot be detected by stuck-at fault techniques.

Stuck-at fault tests, such as scan testing, usually detect SA1 and SA0 faults, but often miss bridged interconnects. This is because the bridged interconnects may have enough resistance to avoid a functional problem and, consequently, avoid detection. I_{DDQ} testing detects bridging faults more efficiently since it is sensitive to high or low resistance bridges and is not dependent on functional failure.

Similarly, gate oxide shorts and tunneling opens can occur when the device is functioning and, therefore, are undetectable through stuck-at fault test detection. However, that same tunneling open or gate oxide short can cause a pair of PMOS and NMOS transistors to conduct simultaneously, thereby increasing quiescent current and consequently being detectable by I_{DDQ} testing.

VLV test is applied to a circuit at operating voltages much lower than its normal operating voltage in order to detect circuit failures. VLV test programs use a set of test patterns along with a predetermined supply voltage to execute the test. The speed of the test vectors and the value of programming voltage are determined from characterizing a known good device from various manufactured wafers. Although VLV can detect more soft defects as supply voltage reduces, it fails to detect interconnect delay failures because the interconnect delay does not scale down when the supply voltage is reduced.

The minimum supply voltage required by VLV test identifies the lowest supply voltage at which the ASIC design can operate properly. In the CMOS process, this low voltage level is normally two to three times higher than the transistor threshold voltages. To determine the lowest operating supply voltage (or reference voltage), VLV test begins with a very low supply voltage such that the device cannot function properly (in voltage failure mode), and then the supply voltage is increased by defined intervals until the device passes the test. This procedure is repeated several times using various known good ASIC devices from the manufacturing wafers and lots to construct minimal supply voltage distribution for analysis to determine the minimum supply voltage level.

For today's deep submicron ASIC designs, the interconnect delay is no longer negligible, and intensive buffer insertion is used during physical design to reduce long interconnect delay. Using I_{DDQ} and VLV testing in conjunction with traditional stuck-at fault testing can improve manufacturing fault detection because of the ability to detect (slow-to-rise and slow-to-fall) defective buffers.

6.7 Built-In-Self-Test

Built-In-Self-Test (BIST) is a useful technique that can improve the quality and reliability of circuits by detecting and correcting faults at an early stage. BIST can also reduce the cost and complexity of testing by eliminating or minimizing the need for external equipment and human intervention.

BIST works by adding extra logic to a circuit that can perform self-testing and self-repair functions. It can be applied to different types of circuits, such as memory, logic, analog, and mixed-signal. Depending on the type of circuit, BIST can use different methods to generate test patterns, apply them to the circuit under test, and compare the outputs with the expected results. Some common methods are:

Logic BIST (LBIST) This method uses a pseudo-random pattern generator, a scan chain, and a signature analyzer to test logic circuits. The pseudo-random pattern generator produces random input patterns that are applied to the scan chain. The scan chain is a series of Flip-Flops that capture the outputs of the logic circuit. The signature analyzer compresses the scan chain outputs into a signature that represents the functionality of the logic circuit. The signature is then compared with a reference signature to determine if the logic is operating properly.

Analog BIST (ABIST) This method uses analog signal generators, ADCs, DACs, and DSP techniques, to test analog circuits. The analog signal generators produce various analog signals that are applied to the analog circuit under test. The ADCs convert the analog outputs of the circuit into digital values. The DACs convert the digital values into analog signals that can be compared with the original inputs. The DSP techniques perform various calculations and analyses on the values to evaluate the performance of the analog circuit.

Mixed-Signal BIST (MSBIST) This method combines ABIST and LBIST techniques to test mixed-signal circuits that have both analog and digital components. MSBIST can use different architectures and strategies to integrate ABIST and LBIST modules and coordinate their operations. For example, MSBIST can use a single controller to control both ABIST and LBIST modules or use separate controllers for each module. MSBIST can also use different modes of operation, such as concurrent mode, where ABIST and LBIST run simultaneously, or sequential mode, where ABIST and LBIST run one after another.

Memory BIST (MBIST)   This method uses a pattern generator, an address generator, and a comparator to test memory arrays. The pattern generator produces various data patterns that are written to and read from the memory cells. The address generator provides the addresses for accessing the memory cells. The comparator checks if the read data matches the written data. If any errors are detected, MBIST can also perform self-repair by replacing the faulty memory cells with redundant ones.

MBIST has been proven to be one of the most comprehensive and effective test methodologies widely used for embedded memory testing. It is very cost-effective because it does not require any external test hardware and requires minimal development effort while allowing, in many cases, to pinpoint the location of defects during testing. However, MBIST has an overhead area penalty and may have some negative timing impacts on the memory performance because of its required logic circuitry. To use MBIST methodology effectively, there are some criteria such as quality and efficiency, which must be considered.

MBIST quality refers to its level of test coverage in detecting stuck-at faults. If the MBIST testing algorithm does not detect many defects, it can influence the quality of the ASIC design and could lead to undesirable engineering and business results. MBIST quality is considered the most important criterion in MBIST methodology.

Efficiency criteria correspond to the area overhead integration of MBIST circuitry in the ASIC design and its test algorithms' run time. Inefficient MBIST can increase the overall ASIC design area and significantly increase the test time for designs that have many embedded memories. The basic MBIST configuration consists of multiplexers, data and address generators, control logic, and comparator logic, as illustrated in Fig. 6.10.

During the MBIST operation, the memory is put into a test mode using multiplexers connected to data, address, and control lines (such as read and write lines), to allow MBIST to exercise the memory independent of the rest of the design. Then, an FSM that resides inside the control logic provides the necessary states and signals for the data and address generator logic to write a test pattern in a March algorithm to the memory cells, read the data back from the memory, and compare it to the original data. If a mismatch occurs during the operation, a flag is set to show that the memory cell under the test has a failure.

It is important to note that although many EDA tools employing MBIST have developed algorithms that are optimized to test a large set of fault types while maintaining reasonable test times, they may not be able to detect faults such as soft defects [4].

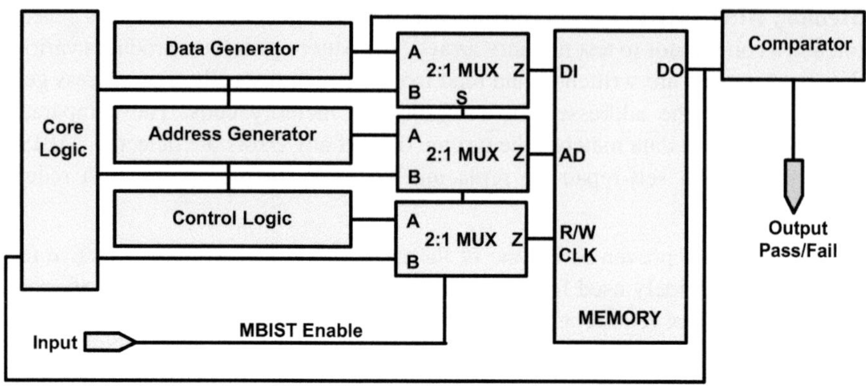

Fig. 6.10 MBIST logical presentation

6.8   Parallel Module Test

The Parallel Module Test (PMT) technique allows test vectors to be applied to blocks or modules within the design and observes their response directly using ASIC primary input and output package pins.

PMT uses multiplexers that can be inserted in a series with a module's normal input and output ports to allow reconfiguration of the interconnect routing of a module from its normal functional routing path to a test access path. Figure 6.11 illustrates PMT configuration of an ASIC design that consists of a memory, hard macro or analog block, and other core logic circuits.

During PMT test, the PMT enabled signal sets the multiplexers such that the module under test can be accessed directly from the ASIC primary input and output. PMT is very useful for memory testing where the area and performance are impacted by MBIST. In addition, PMT can be advantageous from a characterization point of view if an ASIC design has embedded analog modules. By allowing the analog module to be controlled and observed directly via primary input and output pins, PMT reduces test time significantly.

6.9   System Test

As mentioned in Chap. 2, reference board was designed for design RTL development using FPGA to validate the functionality of design and not the design performance due to the speed limitation of FPGA. However, once manufactured ASIC is available, the same reference board can be used with that ASIC with real-time clocking by replacing FPGA with an ASIC device as shown in Fig. 6.12 (based on the reference board in Fig. 2.5 in Chap. 2).

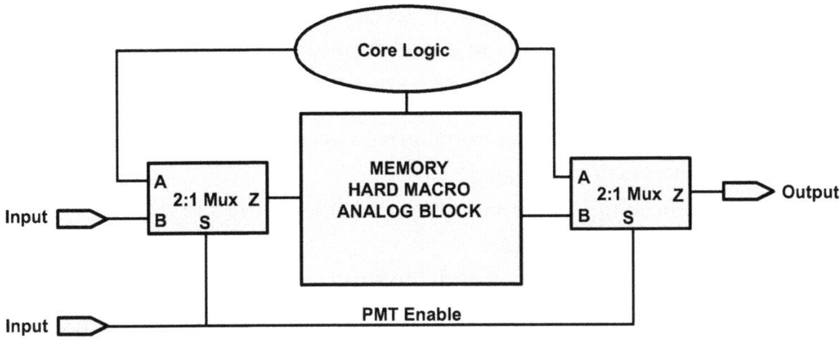

Fig. 6.11   Parallel module test logic

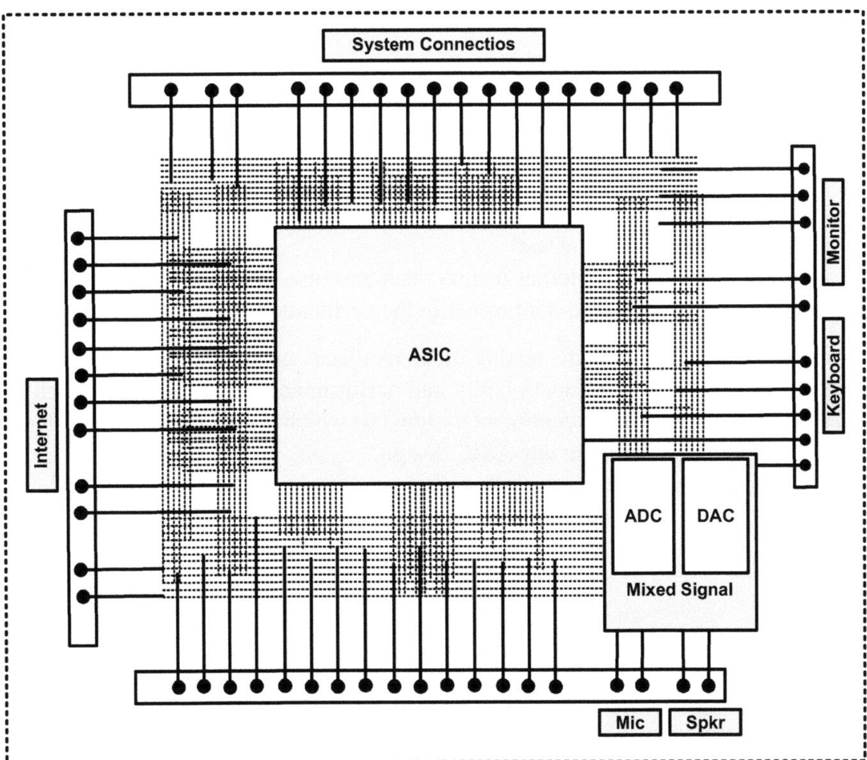

Fig. 6.12   Design reference board using actual ASIC device

ASIC system test using reference board in Lab is a process of validating the functionality and performance of the ASIC chip in a real system environment. It involves connecting the chip to a reference board that has all the necessary components and interfaces for the chip to operate. The reference board can also have sensors, actuators, and other peripherals that are relevant for the target application of the chip. The reference board can run real software and applications that test the chip's features and use cases. The goal is to ensure that the chip meets the customer's expectations and requirements in a true deployment scenario.

Some of the benefits of ASIC system test using reference board in Lab are:

- It can detect and debug any functional errors or performance issues that were not caught during the verification stage.
- It can validate the chip's compatibility and interoperability with other components and devices in the system.
- It can evaluate the chip's power consumption, thermal behavior, and reliability under different operating conditions and stress factors.
- It can demonstrate the chip's functionality and performance to the customer and obtain feedback and approval.

However, there are some challenges of ASIC system test using reference board in Lab as shown below:

- It requires a lot of resources and expertise to design, build, and maintain the reference board and the test environment.
- It can be time-consuming and costly to run the test programs and analyze the test results.
- It can be difficult to reproduce and isolate the root cause of any failures or anomalies that occur during the test.
- It can be affected by external factors such as noise, interference, and environmental variations that are not present in the verification stage.

It should be noted that all the testing methods discussed in this chapter have their own limitations regarding functionality and performance in that they are bound by ATE equipment, with the exception of system test which is not bound by ATE. Thus, the system test is essential in any ASIC design.

6.10   Summary

This chapter briefly discusses the most common ASIC testing techniques such as functional, scan, boundary scan, parametric, current and very low voltage, BIST, memory, and parallel module test.

Today's ASIC designs without DFT can be very expensive and may not meet the high-reliability requirements. It should be recognized that all DFT techniques require design trade-offs such as area, design cycle time, or extraction processing steps. Economic feasibility of design trade-offs must be carefully evaluated in relation to test development cycle time, test time reduction, as well as the level of fault coverage achieved.

The functional test section outlines the importance of proper circuit initialization, use of multiplexers, and basic DFT requirements. Functional testing can involve either manual or automated methods. Although manual testing methods are appropriate for some designs, they are time-consuming and inefficient for complex

ASIC requiring shorter test development cycles. Automated testing, on the other hand, can allow increased test coverage and shorter development cycles that may not be possible with manual testing.

In the scan and boundary scan test section, the concept of such testing techniques and their operation under testing conditions is discussed. Although scan and boundary scan improve the stuck-at coverage, it may not be sufficient to cover many timing-based defects. To address this inefficiency, scan-based ATPG solutions for at-speed testing are required. Having such a solution not only ensures high test coverage of detecting timing-based defects, but it also reduces testing time.

In the parametric testing section, the concepts of AC and DC testing have been covered. These types of tests are performed after wafer manufacturing is completed. During parametric testing, various technological and process parameters are measured and compared against specifications to ensure proper device manufacturing.

In the current and very low-level voltage test section, we outlined alternative testing strategies that can detect additional defects such as tunneling opens that cannot be detected by functional or at-speed testing. These types of process-induced defects are major types of defects for submicron process, and we need to be able to detect such defects to improve the device reliability.

The memory and parallel test section discussed the advantage of these tests for DFT improvement. In addition, we covered automatic test generation for memories. From a physical design point of view, having this testing capability allows one to reduce the test development cycle time especially if the design contains many memories.

References

1. E.B. Eichelberger and T.W. Williams, "A logic design structure for LSI testability", *Journal of Design Automation and Fault Tolerant Computing*, pp. 165–78, 1978
2. ANSI/IEEE Standard 1149.1-1990, *IEEE Standard Test Access Port and Boundary-Scan Architecture*, IEEE Standards Board, New York, N.Y, May 1990
3. P. Franco "Testing Digital Circuits for Timing Failures by Output Waveform Analysis", *Center for Reliable Computing Technical Report*, No. 94-9 Stanford University, 1994
4. Khosrow Golshan, *Physical Design Essentials, an ASIC Design Implementation Perspective*, Springer Business Media, 2007

Chapter 7
ASIC Qualification

Success is stumbling from failure to failure with no loss of enthusiasm.
–Winston Churchill

The ASIC chip qualification process is a series of tests and procedures that ensure the quality and reliability of an ASIC. An ASIC is a custom-made chip that is designed for a specific purpose, such as automotive, graphic processing, mining cryptocurrencies, processing audio or video signals, or performing complex calculations such as generative artificial intelligence. The qualification process verifies that the ASIC meets the functional, performance, and environmental requirements of the target application and market.

It's important to check the required quality level by conducting customer interviews. Some customers will have stringent quality requirements; therefore, it may require conducting longer and more comprehensive qualification tests. It's recommended a minimal qualification test using Multi-Project Wafer (MPW) prototypes. MPW service is a cost-effective way of prototyping an ASIC by sharing the same silicon among multiple customers. Each customer can design their own ASIC and submit it to an MPW service provider, who will then fabricate the wafer and deliver the ASICs to the customer. This is also true for ASIC design companies to have their own MPW and submit that the processing foundries. MPW service reduces the cost and time of ASIC development by utilizing the unused space on the wafer. Using MPW allows ASIC vendors using MPW to ensure the major qualification tests are successful before the full mask-set for production tape-out. Figure 7.1 shows the concept MPW using different ASIC designs (i.e., ASIC A, B, C, and D), and Fig. 7.2 shows the production of given ASIC design (e.g., ASIC A).

The qualification process typically involves the following steps:

Technology qualification This step evaluates the fabrication process and the technology platform of the ASIC. It involves testing the basic electrical parameters, such as voltage, current, resistance, capacitance, and inductance, of the ASIC. It also involves testing the physical characteristics, such as dimensions, layout, and

© The Author(s), under exclusive license to Springer Nature Switzerland AG 2024
K. Golshan, *ASIC Design Implementation Process*,
https://doi.org/10.1007/978-3-031-58653-8_7

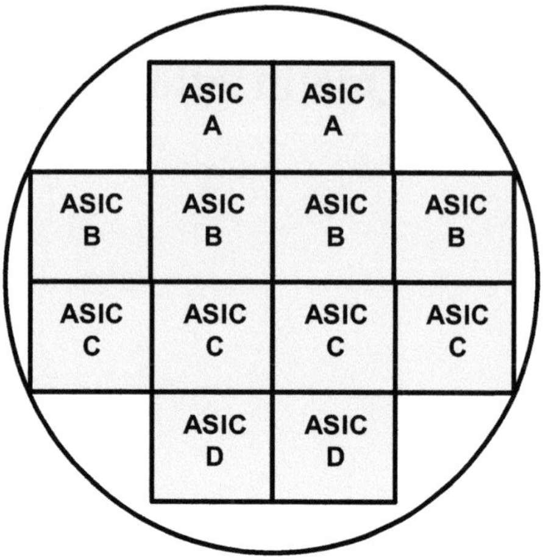

Fig. 7.1 Conceptual MPW
ASICs

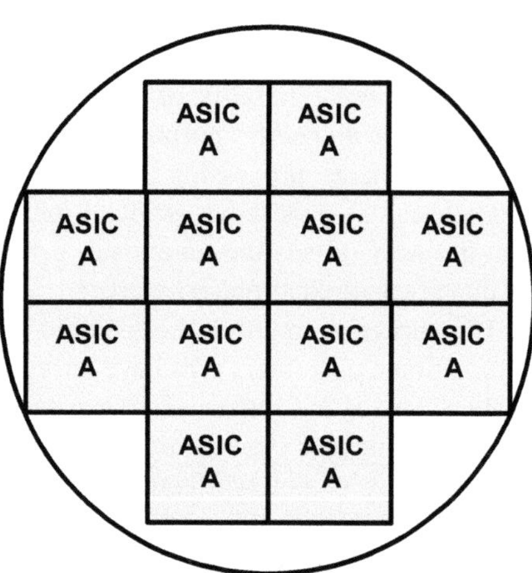

Fig. 7.2 ASIC A's
production wafer

packaging, of the ASIC. The goal of this step is to ensure that the ASIC is manufactured according to the design specifications and industry standards.

Part number qualification This step evaluates the functionality and performance of the ASIC. It involves testing the input and output pins, the logic gates, the memory cells, and the arithmetic units of the ASIC. It also involves testing the timing, power consumption, and thermal characteristics of the ASIC. The goal of this step is to ensure that the ASIC performs as expected and meets the functional requirements of the target application.

Supplier process survey This step evaluates the quality and reliability of the supplier of the ASIC. It involves auditing the supplier's facilities, equipment, personnel, documentation, and procedures. It also involves reviewing the supplier's quality control, defect analysis, failure analysis, and corrective action process. The goal of this step is to ensure that the supplier has adequate capabilities and resources to produce high-quality and reliable ASICs.

Part construction analysis   This step evaluates the internal structure and composition of the ASIC. It involves using various techniques, such as scanning electron microscopy (SEM), Energy-Dispersive X-ray spectroscopy (EDX), Focused Ion Beam (FIB), and transmission electron microscopy (TEM), to examine the layers, materials, interconnections, and defects of the ASIC. The goal of this step is to ensure that the ASIC is free of any physical or chemical anomalies that could affect its functionality or reliability.

Failure rate projection   This step evaluates the expected lifetime and failure rate of the ASIC. It involves using various models, such as MIL-HDBK-217F or Telcordia SR-332, to estimate the Mean Time Between Failures (MTBF) or Mean Time to Failure (MTTF) of the ASIC. It also involves using various methods, such as Accelerated Life Testing (ALT) or Highly Accelerated Life Testing (HALT), to simulate the stress conditions that could cause failures in the ASIC. The goal of this step is to ensure that the ASIC meets the reliability requirements of the target application and market. The qualification process may vary depending on the type, complexity, and purpose of the ASIC. Some common tests that are used in the qualification process are [1]:

- ESD test (HBM and CDM)
- Latch-up test
- Wafer Acceptance Test (WAT)

7.1   Electro-Static Discharge Test

In designing ASICs, one area of concern is how they are protected from outside electrical sources. While most systems are geared toward power overloads, one source that may cause considerable damage is the ESD that comes from the human body. To help test for ESD, the Human Body Model (HBM) is used.

The HBM is the model used to characterize the susceptibility of electronic devices that might be subject to damage from ESD. In the United State, there is a military standard of HBM used (MIL-STD-883, Method 3015.9) which is now common across all testing for ESD. This national standard makes it simpler for testing to be performed and provides a singular model so that all devices can be judged on their resistance to ESD. There is another standard used internationally called JEDEC JS-0001 [2].

The way HBM test is performed is that under both national and international standards, the HBM is represented by a 100-pF capacitor combined with a

1500-ohm discharge resistance. Under the test, several kilo-volts are charged into the capacitor which usually include the following levels:

- 2 kV
- 4 kV
- 6 kV
- 8 kV

The electricity is discharged through the resistor which is connected into the series of the IC, and the results are observed. The HBM provides a standard by which the ESD can be measured, as shown in Fig. 7.3, and the results are seen so that any necessary changes can be made to the ASIC for additional protection.

In addition to HBM, Charged Device Model (CDM) must be used to test for ESD. Although superficially like the HBM, the CDM is different and helps simulate the ESD caused by two different sources: the triboelectric effect (which may cause damage to an electronic device directly) or electrostatic induction (which may cause damage indirectly to an electronic device). This static charge may be building up inside a part or area and may discharge under the certain conditions.

The CDM test will simulate the situations where devices are handled in manufacturing locations and areas, as shown in Fig. 7.4, such as when they slide down tubes or other surfaces. A standard CDM ESD test will measure the characteristics of the waveforms when an external ground touches the ASIC pin of the device which is charged as shown in Fig. 7.3. This buildup is then discharged from the device and into the ground. While the device is kept in a position, the discharge will be measured, and the effect recorded. CDM are higher and stronger than HBM discharges because there are no inherent limits and will reveal any vulnerabilities of the electronic device caused by this form of discharge. ESD test is the process of exposing ASIC/ICs' IO pins to transient high voltage to ensure ESD protection.

In addition to HBM and CMD, Machine Model (MM) is a type of ESD test for ASIC qualification. It simulates the discharge that occurs when a charged machine or tool touches an ASIC pin. The MM test applies a high voltage pulse between the tested pin and the ground and measures the current waveform and the failure threshold. The MM test is one of the standard methods to evaluate the robustness of the

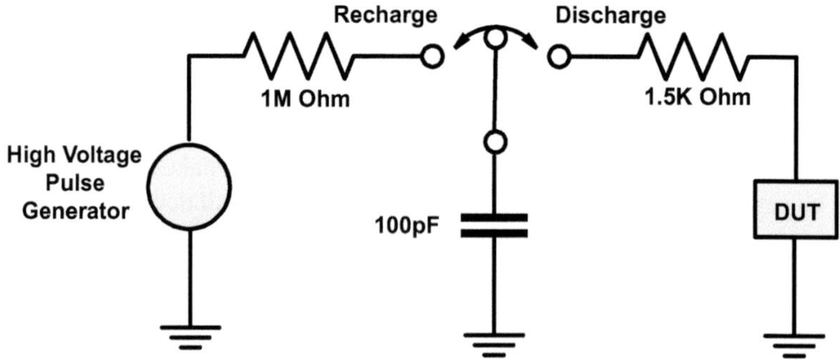

Fig. 7.3 An example of HBD ESD testing

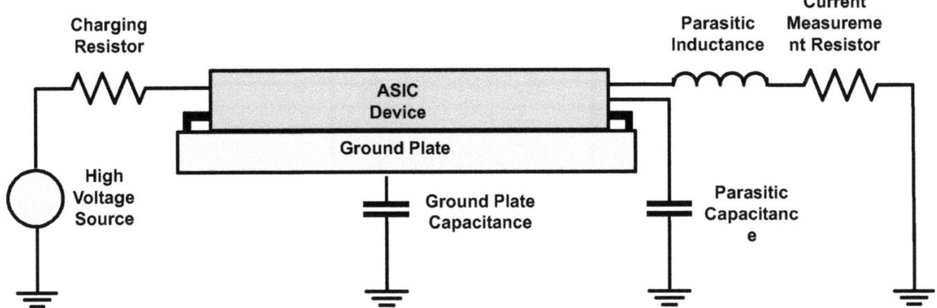

Fig. 7.4 An example of CMD ESD testing

ASIC against ESD damage. The following is the recommended applied voltages for such tests:

- HBM: 2KV
- CDM: 0.5KV
- MM: 0.2KV

It is important to note that one of the most critical elements of ASIC design is the design of the I/O pads. I/O pad structures require the deepest circuit design expertise and most detailed understanding of the process that is being used to fabricate an ASIC design. I/O pad circuits translate the signal levels used in the ASIC core to the signal levels used outside the ASIC. Additionally, the I/O pad circuits clamp signals to the power and ground rails to limit the voltage at the external connection to the ASIC I/O pad. This clamping reduces signal overshoot and prevents ESD damage .

The most effective way to prevent ESD is to design power and ground pads to provide connections to the various ASIC power and ground buses. The metal connections from the pad to the power or ground bus within the power and ground pad or to the ASIC core are made as wide as possible and on as many metal layers as practical to minimize their resistance and maximize the current-carrying capacity. Because no other circuitry is required, the area in power and ground pads is often used for special ESD clamping circuitry that is associated only with power buses and not with any signal pin. Figure 7.5 shows a basic I/O and core power pad with built-in ESD protection circuitry.

In addition to the ESD protection circuitry within the I/O pad, there are other clamping devices coupling the various power and ground buses to each other. These clamping devices may be simple diodes connecting two supplies having the same voltage—such as digital ground to an isolated analog ground.

ESD events can lead to the failure of poorly designed I/O pads. As mentioned earlier, this failure mechanism is primarily thermally driven and is mainly caused by three physical factors—human body, charge device, and/or mechanical handling of an ASIC device. It important to note that future advancements in process technologies that will produce thinner oxide layers, narrower line widths on conducting layers, and shallower junction transistors will make ASIC devices more susceptible

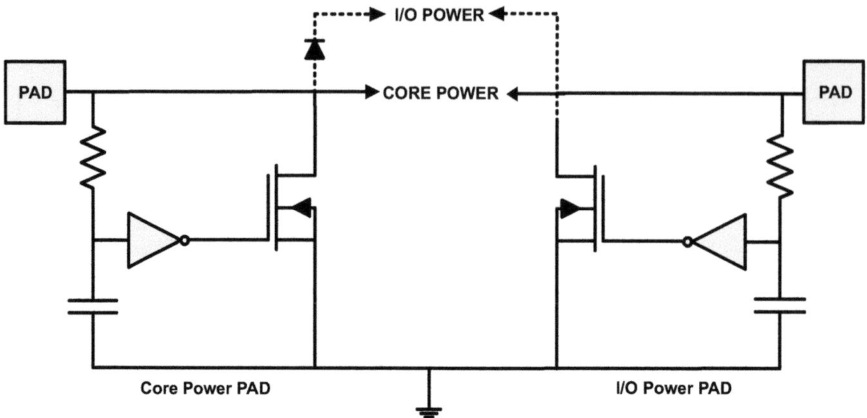

Fig. 7.5 Basic core and I/O power PAD

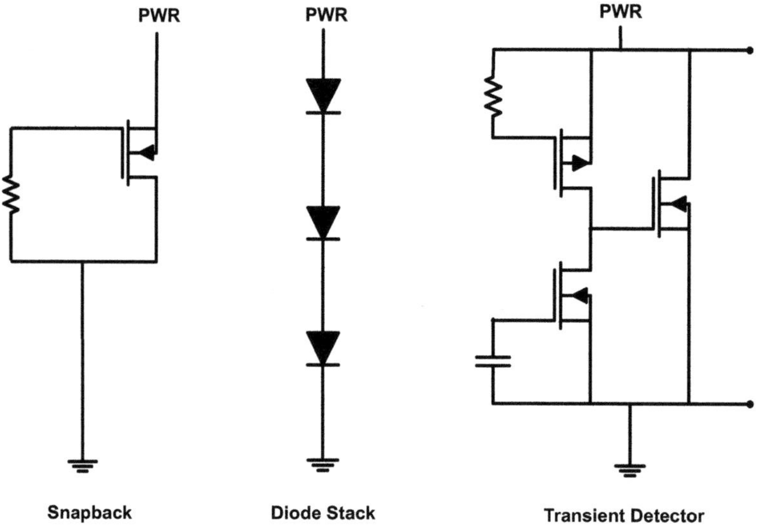

Fig. 7.6 Power supply ESD clamping

to ESD damage. To prevent I/O from ESD damage, CLAMP design between power and ground can be utilized.

For CLAMPs between power and ground, as shown in Fig. 7.6, a single transistor may be used as a snapback device, or it may be used for more complex circuits such as diode stacks (for low-voltage supplies) or large transistors with transient detector circuits.

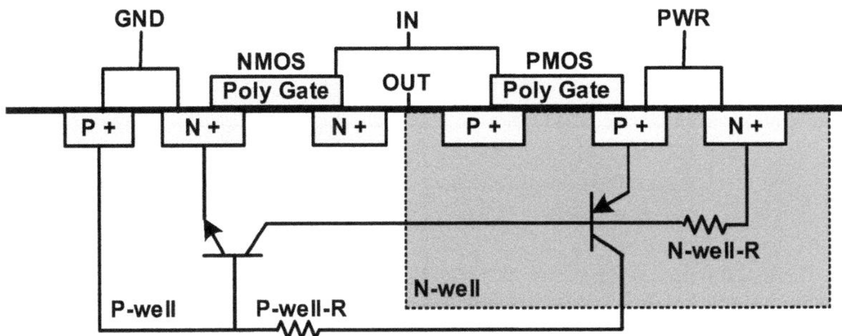

Fig. 7.7 Typical CMOS inverter circuit

7.2   Latch-up Test

Latch-up is a functional chip failure associated with excessive current going through the manufactured ASIC, caused by weak circuit design. In some cases, latch-up can be a temporary condition that can be resolved by power cycle, but unfortunately, it can also cause a fatal ASIC failure.

To help understand the latch-up phenomena, Fig. 7.7 shows a typical CMOS inverter circuit. Notice the two bipolar transistors, NPN and PNP, and their connection to PWR and GND supply rails (also known as thyristor-like structures). The two transistors are protected by resistors, but if examined more closely, the latch-up is often caused by the interaction of the parasitic thyristor-like structures inherent in CMOS technology. These structures are essentially Silicon-Controlled Rectifiers (SCRs), and if a voltage spike occurs at the inputs of the CMOS device, it can trigger these parasitic SCRs, causing latch-up.

In a latch-up conduction, the current flows from PWR to GND directly via the two transistors, P-N-P and N-P-N, causing the dangerous condition of a short circuit. The resistors are bypassed, and, thus, excessive current flows from PWR to ground as shown in Fig. 7.8.

In addition to preventing the internal ASIC standard cells from latch-up, it is also important when designing I/O pads to insure their immunity to the latch-up problem. This problem is more severe in I/O pads than in standard cells. This is because signal overshoot causes the transistor drain-body diode to become forward biased, resulting in current flow through the substrate.

The latch-up phenomenon is well understood and is inherent to bulk CMOS process. The result of this effect is the shortening between power and ground lines that can lead to ASIC self-destruction and system power failure. One of the most common practices during I/O pad circuit design is to separate NMOS and PMOS transistors and encircle them with the appropriate well tie and guard ring. Figures 7.9 and 7.10 show the N+ and P+ well ties and guard rings that surround the PMOS and NMOS transistors. The output driver transistors that are directly connected to the PAD are each isolated within their own rings.

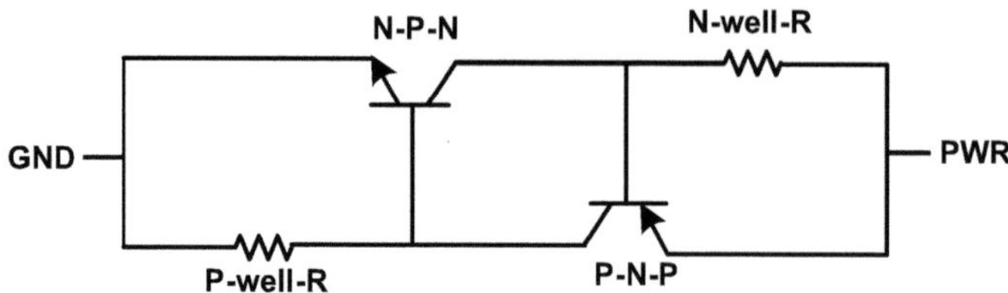

Fig. 7.8 Electrical presentation between N-P-N and P-N-P transistors

Fig. 7.9 P+ well ties and guard ring for NMOS transistor PAD

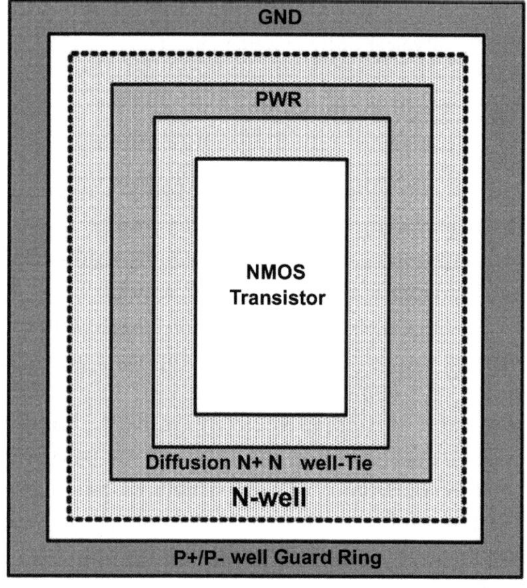

The pre-driver and input receiver transistors are grouped by PMOS and NMOS, and each type is isolated from the other as well as the output driver transistors. If an NWELL is tied to a signal instead of to power (such as an NWELL resistor), then it should be treated as N-type diffusion during physical verification (e.g., latch-up rule verification).

The latch-up test is a series of attempts that trigger the SCR structure within the CMOS IC, while the relevant pins are monitored for overcurrent behavior. It's recommended to take the very first samples from the engineering lot or MPW run and send them to a latch-up testing lab. The lab will apply the maximum possible supply power and then inject current to the chip inputs and outputs while measuring if a latch-up occurs by monitoring the supply current.

There are more qualification tests related to quality and reliability tests, such as stress tests. Some of the common stress tests used are as follows:

- HTOL (High Temperature Operating Lifetime) test
- HTS (High Temperature Storage) test
- HAST (High Accelerated Stress Test)

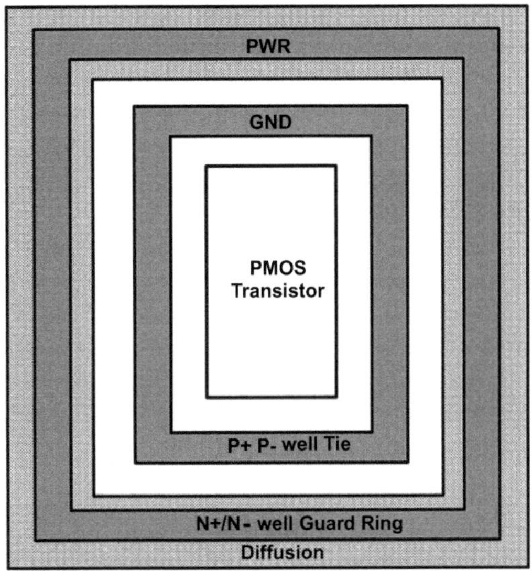

Fig. 7.10 N+ well ties and guard ring for PMOS transistor PAD

- TC (Temperature Cycling) test
- Preconditioning Test

HTOL is a stress test defined by JEDEC to predict the reliability and operating life of IC products. HTOL exposes the IC to extreme temperature conditions. Test results are then used to predict the long-term failure rate of the IC. Almost every ASIC that is going to production must go through HTOL testing.

An HTOL test, according to JEDEC Standard (JESD22-108 and JESD S85), is performed in an oven of 125C degrees, while the ICs are activated with dynamic signals and the VDD pins, with max voltage. The test duration is 1000 hours with zero failure for acceptance criteria. However, in the oven, the ICs are, ideally, placed in sockets and located on a PCB. Both sockets and burn-in-board must be able to withstand the high temperature during the test.

After the HTOL stress test is completed, the ICs must go through electrical screening to determine how many devices passed or failed the stress test. The JEDEC requirements define zero failures as an acceptable criterion, with test results defined in terms of Failures in Time (FIT). One FIT symbolizes one failure in 109 device hours.

HAST involves the effects of humidity and temperature on an IC or ASIC. The HAST is designed to test the package of the ASIC under extreme humidity and temperature conditions. Devices that pass such tests will be able to withstand the normal rigors of temperature and humidity in most environments.

HAST is sometimes called the pressure-cooker test, because it simulates the conditions found inside a typical household pressure cooker. By increasing the amount of water vapor pressure inside a test chamber while also increasing the temperature, this provides an excellent test of the inherent strength, design, and protection of the electronic components inside. The testing involves increasing the pressure which helps to force water vapor inside the device.

Compared to the former standard of high temperature/high humidity, the HAST will cause more damage to the components due to how the moisture helps accelerate the corrosive effects of the water vapor. There will also be more damage to the insulation as well, causing fast deterioration, especially in devices that are not up to the HAST standards.

This type of test is mostly done on components that are plastic-sealed, followed by an evaluation of the results. For the most part, the temperatures that are used reach at least 212F degrees or that of boiling water. This provides for a full state of water vapor to be present in the air and pressurized for maximum effect.

The HAST test is performed based on JEDEC standard JESD22-A110 (biased HAST) and JESDA118 (unbiased HAST). With the following conditions:

- 130 °C/85% RH/33.3 PSIA
- 110 °C/85% RH.17.7 PSIA
- Test duration is 96 or 264 hours

There are recommended environmental conditions for storage of ASICs or ICs which allow them to perform at proper functionally. However, it is well known that storage may not always be close to the recommended standard. This is where the HTSL test comes into play, testing the parameters of the device, so that it can be determined what damage will result from being stored in less-than-ideal conditions.

The HTSL test is in response to how many businesses and consumers may put away an electronic device for a considerable length of time in conditions that may include exposure to high temperature.

This type of test is used to screen, monitor, qualify, or evaluate all ASICs, which means they have no moving parts. The test itself is used to determine the overall effects of temperature and the passage of time for devices that are stored. This means that the device is tested to see how it reacts to being stored in high- or low-temperature environments along with the time it may spend in storage before being used. This includes data retention failure mechanisms or non-volatile memory devices.

JEDEC has a few documents describing the different tests. This information can be found under www.jedec.org.

7.3 Wafer Acceptance Test

Wafer Acceptance Testing, or WAT, is commonly referred to as element evaluation of qualification. It is a sampling method that is used to determine if the manufactured ASIC device will meet the design specifications. WAT provides an economical means for both the semiconductor manufacturer and ASIC providers to separate wafers that are likely to have higher yield from those that will not.

There are two important aspects of using WAT data. One involves the evaluation of ASIC manufacturer process, and the second monitors the process during ASIC manufacturing. For process evaluation, WAT data supplied by the semiconductor manufacturer is compared with various process parameters such as transistor characteristics, resistance, capacitance, and noise figures. This allows one to understand the correlation between the actual process and models, as well as to make sure the process meets ASIC design physical and electrical specifications.

Furthermore, the ASIC design team should evaluate the semiconductor process capabilities for wafer acceptance. Often wafer acceptance is accomplished by screening several simultaneously fabricated wafers containing test chips to predict if a set of wafers (or lot) will yield an adequate and acceptable percentage of good parts or die.

By measuring PMOS and NMOS parameters such as delay and comparing those with the model, one can observe if the process behavior is in line with the model or not. PMOS and NMOS transistors are simulated for various conditions such as fast-PMOS/fast-NMOS, fast-PMOS/slow-NMOS, slow-PMOS/fast-NMOS, and slow-PMOS/slow-NMOS using models (e.g., SPICE). The results (as indicated by the indices of the trapezoid-shaped region shown in Fig. 7.11) are plotted along with measured WAT data.

In Fig. 7.11, the data points inside the acceptable region are an indication of process and model alignment. However, the data residing outside the acceptable region represents mismatches between the process and the models. From a production point of view, these types of mismatches between process and models need to be resolved. One resolution is to extend the acceptable region in Fig. 7.11 so that all the outlying data is covered by adjusting the PMOS and NMOS transistor models. This approach may not be desirable, because it requires recharacterization of the entire spectrum of ASIC libraries (standard cells, memories, and intellectual properties) and can have a negative impact on the existing ASIC design. Another solution to the problem is to modify the current process to match existing transistor models. This method is preferred because there is no model or library changes required.

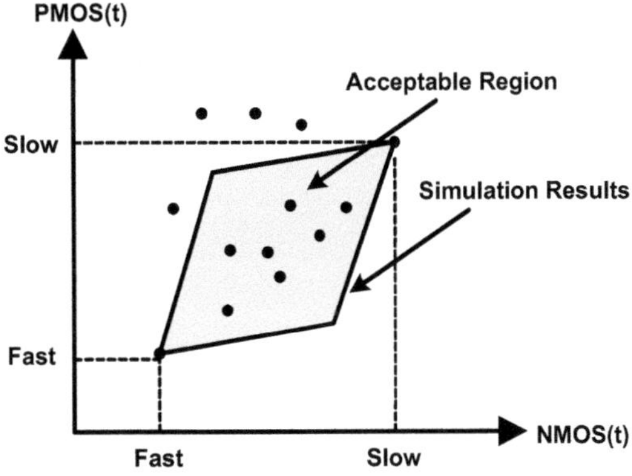

Fig. 7.11 Transistor simulation results versus WAT data

WAT data is generated through visual inspection or parametric testing. The most common process parameters are as follows:

- Critical dimension (e.g., polysilicon length)
- Oxide thickness
- Thin film thickness
- Sheet resistivity
- Current and voltage shift
- Current leakage
- Mask alignment
- Particle defects
- Under-/over-etching
- Capacitance/resistance mismatch
- Diode characteristics (PMOS and NMOS performance)

After the ASIC is submitted to the manufacturer for processing, two physical rings will be added to the actual ASIC device before masks are produced. These rings are called the seal ring and scribe line as shown in Fig. 7.12.

The seal ring joins the package case header and substrate to their covers to prevent penetration of moisture into the package. The scribe line is used to separate a processed wafer into individual die. Scribe lines, or narrow channels between individual die, are mechanically weakened by scratching with a scribe (diamond tip), sawing with a diamond blade, or burning with a laser. The scribe lines are then mechanically stressed and broken apart along the scribe lines, thus separating the individual die.

To save silicon area, almost all ASIC manufacturers use scribe lines as a vehicle to add their process-monitoring test structures to collect WAT data during the device fabrication process. In recent years, ASIC and semiconductor industries have

Fig. 7.12 ASIC seal ring and scribe line

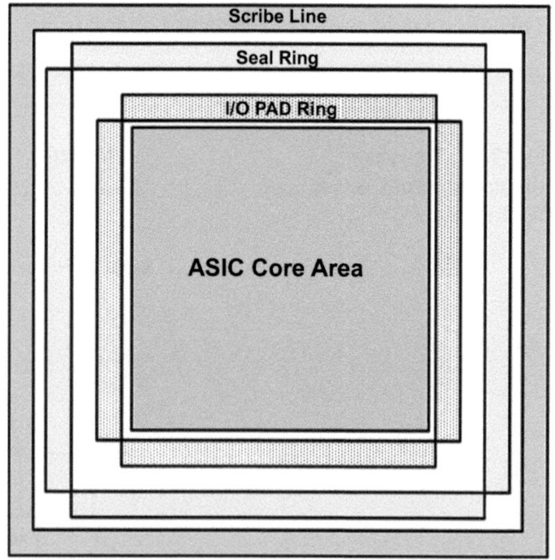

intensified their attention to yield issues to meet the challenges of manufacturing complex ASIC (at 130 nm technology and below); thus, monitoring WAT data during design production is key to insuring high yield.

7.4   Summary

This chapter discusses an ASIC qualification test. In this chapter, an overview of ESD is discussed that includes several tests such as HBM, CMD, and MM. In addition, the sources of ESD failure and design techniques that prevent ESD damage for IO and standard cells are provided.

In addition, the concept of WAT as element evaluation of qualification is discussed. WAT is a sampling method that is used to determine if the manufactured ASIC device will meet the design specifications. WAT provides an economical means for both the semiconductor manufacturer and ASIC providers to separate wafers that are likely to have higher yield from those that will not.

The role of ASIC manufacturers in the qualification process is also discussed. They add the seal ring which joins the package case header and substrates to their covers to prevent penetration of moisture into the package. They also use a scribe line to separate a processed wafer into individual die.

References

1. Khosrow Golshan, *Physical Design Essentials, an ASIC Design Implementation Perspective*, Springer Business Media, 2007
2. JEDEC Solid State Technology Association, *ANSI/ESDA/JEDEC JS-00*, ESD Association and JEDEC Solid State Association, Arlington VA. USA, 2012

Index

MIX
Papier aus verantwortungsvollen Quellen
Paper from responsible sources
FSC® C105338

If you have any concerns about our products,
you can contact us on
ProductSafety@springernature.com

In case Publisher is established outside the EU,
the EU authorized representative is:
Springer Nature Customer Service Center GmbH
Europaplatz 3, 69115 Heidelberg, Germany

Printed by Libri Plureos GmbH
in Hamburg, Germany